Tools Teach
An Iconography of
American Hand Tools

Hand Tools in History Series

- Volume 6: Steel- and Toolmaking Strategies and Techniques before 1870
- Volume 7: Art of the Edge Tool: The Ferrous Metallurgy of New England Shipsmiths and Toolmakers
- Volume 8: The Classic Period of American Toolmaking, 1827-1930
- Volume 9: An Archaeology of Tools: The Tool Collections of the Davistown Museum
- Volume 10: Registry of Maine Toolmakers
- Volume 11: Handbook for Ironmongers: A Glossary of Ferrous Metallurgy Terms: A Voyage through the Labyrinth of Steel- and Toolmaking Strategies and Techniques 2000 BCE to 1950
- Volume 13: Tools Teach: An Iconography of American Hand Tools

Tools Teach
An Iconography of
American Hand Tools

H. G. Brack

Davistown Museum Publication Series
Volume 13

ISBN 978-0-9829951-8-1

Copyright © 2013 by H. G. Brack
ISBN 13: 978-0-9829951-8-1
ISBN 10: 0982995180
Davistown Museum

First Edition; Second Printing

Photography by Sett Balise

Cover illustration by Sett Balise includes the following tools:
Drawshave made by I. Pope, 913108T51
Dowel pointer, 22311T11
Inclinometer level made by Davis Level & Tool Co., 102501T1
Expansion bit patented by L. H. Gibbs, 090508T6
Socket chisel, 121805T6
Bedrock No. 2 smooth plane made by Stanley Tool Company, 100400T2
Molders' hand tool, 102112T3
Caulking iron made by T. Laughlin Co. of Portland, ME, TCX1005
T-handle wood threading tap, 102212T2
Silversmiths' hammer head made by Warner & Noble of Middletown, CT, 123012T3
Wire gauge made by Morse Twist Drill & Machine Co. of New Bedford MA, 10910T5
Surface gauge made by Veikko Arne Oby of Whitinsville, MA, 21201T12
No. 77 adjustable nut wrench made by W. J. Ladd of NY, 31212T6
Plumb bob made by Stanley Tool Co., 71401T18
Gouge made by Buck Brothers of Millbury, MA, 42904T2.

Back cover illustration: Wooden plow plane made by I. Holmes, courtesy of Rick Floyd.

This publication was made possible by a donation from Barker Steel LLC.

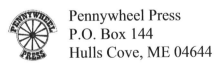
Pennywheel Press
P.O. Box 144
Hulls Cove, ME 04644

Preface

Davistown Museum *Hand Tools in History*

One of the primary missions of the Davistown Museum is the recovery, preservation, interpretation, and display of the hand tools of the maritime culture of Maine and New England (1607-1900). The *Hand Tools in History* series, sponsored by the museum's Center for the Study of Early Tools, plays a vital role in achieving the museum mission by documenting and interpreting the history, science, and art of toolmaking. The Davistown Museum combines the *Hand Tools in History* publication series, its exhibition of hand tools, and bibliographic, library, and website resources to construct an historical overview of steel- and toolmaking strategies and techniques used by the edge toolmakers of New England's Wooden Age. Included in this overview are the roots of these strategies and techniques in the early Iron Age, their relationship with modern steelmaking technologies, and their culmination in the florescence of American hand tool manufacturing in the last half of the 19th century.

Background

During over 40 years of searching for New England's old woodworking tools for his Jonesport Wood Company stores, curator and series author H. G. Skip Brack collected a wide variety of different tool forms with numerous variations in metallurgical composition, many signed by their makers. The recurrent discovery of forge welded tools made in the 18th and 19th centuries provided the impetus for founding the museum and then researching and writing the *Hand Tools in History* publications. In studying the tools in the museum collection, Brack found that, in many cases, the tools seemed to contradict the popularly held belief that all shipwrights' tools and other edge tools used before the Civil War originated from Sheffield and other English tool-producing centers. In many cases, the tools that he recovered from New England tool chests and collections dating from before 1860 appeared to be American-made rather than imported from English tool-producing centers. Brack's observations and the questions that arose from them led him to research the topic and then to share his findings in the *Hand Tools in History* series.

Hand Tools in History Publications

- Volume 6: *Steel- and Toolmaking Strategies and Techniques before 1870* explores ancient and early modern steel- and toolmaking strategies and techniques, including those of early Iron Age, Roman, medieval, and Renaissance metallurgists and toolmakers. Also reviewed are the technological innovations of the Industrial Revolution, the contributions of the English industrial revolutionaries to the evolution of the factory system of mass production with interchangeable parts, and the development of bulk steelmaking processes and alloy steel technologies in the latter half of the 19th century. Many of these technologies play a role in the florescence of American ironmongers and toolmakers in the 18th and 19th centuries. Author H. G. Skip Brack cites archaeometallurgists such as Barraclough, Tylecote, Tweedle, Smith, Wertime, Wayman, and many others as useful guides for a journey through the pyrotechnics of ancient and modern metallurgy. Volume 6 includes an extensive bibliography of resources pertaining to steel- and toolmaking techniques from the early Bronze Age to the beginning of bulk-processed steel production after 1870.

- Volume 7: *Art of the Edge Tool: The Ferrous Metallurgy of New England Shipsmiths and Toolmakers* explores the evolution of tool- and steelmaking techniques by New England's shipsmiths and edge toolmakers from 1607-1882. This volume uses the construction of Maine's first ship, the pinnace *Virginia*, at Fort St. George on the Kennebec River in Maine (1607-1608), as the iconic beginning of a critically important component of Colonial and early American history. While there were hundreds of small shallops and pinnaces built in North and South America by French, English, Spanish, and other explorers before 1607, the construction of the *Virginia* symbolizes the very beginning of New England's three centuries of wooden shipbuilding. This volume explores the links between the construction of the *Virginia* and the later flowering of the Colonial iron industry; the relationship of 17th, 18th, and 19th century edge toolmaking techniques to the steelmaking strategies of the Renaissance; and the roots of America's indigenous iron industry in the bog iron deposits of southeastern Massachusetts and the many forges and furnaces that were built there in the early Colonial period. It explores and explains this milieu, which forms the context for the productivity of New England's many shipsmiths and edge toolmakers, including the final flowering of shipbuilding in Maine in the 19th century. Also included is a bibliography of sources cited in the text.

- Volume 8: *The Classic Period of American Toolmaking 1827-1930* considers the wide variety of toolmaking industries that arose after the Colonial period and its robust tradition of edge toolmaking. It discusses the origins of the florescence of American toolmaking not only in English and continental traditions, which produced gorgeous hand tools in the 18th and 19th centuries, but also in the poorly documented and often unacknowledged work of New England shipsmiths, blacksmiths, and toolmakers. This volume explicates the success of the innovative American factory system, illustrated by an ever-expanding repertoire of iron- and steelmaking strategies and the widening variety of tools produced by this factory system. It traces the vigorous growth of an American hand toolmaking industry that was based on a rapidly expanding economy, the rich natural resources of North America, and continuous westward expansion until the late 19th century. It also includes a company by company synopsis of America's most important edge toolmakers working before 1900, an extensive bibliography of sources that deal with the Industrial Revolution in America, special topic bibliographies on a variety of trades, and a timeline of the most important developments in this toolmaking florescence.

- Volume 9: *An Archaeology of Tools* contains the ever-expanding list of tools in the Davistown Museum collection, which includes important tools from many sources. The tools in the museum exhibition and school loan program that are listed in Volume 9 serve as a primary resource for information about the diversity of tool- and steelmaking strategies and techniques and the locations of manufacturers of the tools used by American artisans from the Colonial period until the late 19th century.

- Volume 10: *Registry of Maine Toolmakers* fulfills an important part of the mission of the Center for the Study of Early Tools, i.e. the documentation of the Maine toolmakers and planemakers working in Maine. It includes an introductory essay on the history and social context of toolmaking in Maine; an annotated list of Maine toolmakers; a bibliography of sources of information on Maine toolmakers; and appendices on shipbuilding in Maine, the metallurgy of edge tools in the museum collection, woodworking tools of the 17th and 18th centuries, and a listing of important New England and Canadian edge toolmakers working outside of Maine. This registry is available on the Davistown Museum website and can be accessed by those wishing to research the history of Maine tools in their possession. The author greatly appreciates receiving information about as yet undocumented Maine toolmakers working before 1900.

- Volume 11: *Handbook for Ironmongers: A Glossary of Ferrous Metallurgy Terms* provides definitions pertinent to the survey of the history of ferrous metallurgy in the preceding five volumes of the *Hand Tools in History* series. The glossary defines terminology relevant to the origins and history of ferrous metallurgy, ranging from ancient metallurgical techniques to the later developments in iron and steel production in America. It also contains definitions of modern steelmaking techniques and recent research on topics such as powdered metallurgy, high resolution electron microscopy, and superplasticity. It also defines terms pertaining to the growth and uncontrolled emissions of a pyrotechnic society that manufactured the hand tools that built the machines that now produce biomass-derived consumer products and their toxic chemical byproducts. It is followed by relevant appendices, a bibliography listing sources used to compile this glossary, and a general bibliography on metallurgy. The author also acknowledges and discusses issues of language and the interpretation of terminology used by ironworkers over a period of centuries. A compilation of the many definitions related to iron and steel and their changing meanings is an important component of our survey of the history of the steel- and toolmaking strategies and techniques and the relationship of these traditions to the accomplishments of New England shipsmiths and their offspring, the edge toolmakers who made shipbuilding tools.

- Volume 13 in the *Hand Tools in History* series explores the iconography (imagery) of early American hand tools as they evolve into the Industrial Revolution's increased diversity of tool forms. The hand tools illustrated in this volume were selected from the Davistown Museum collection, most of which are cataloged in *An Archaeology of Tools* (Volume 9 in *Hand Tools in History*), and from those acquired and often sold by Liberty Tool Company and affiliated stores, collected during 40+ years of "tool picking." Also included are important tools from the private collections of Liberty Tool Company customers and Davistown Museum supporters. Beginning with tools as simple machines, reviews are provided of the metallurgy and tools used by the multitasking blacksmith, shipsmith, and other early American artisans of the Wooden Age. The development of machine-made tools and the wide variety of tool forms that characterize the American factory system of tool production are also explored. The text includes over 800 photographs and illustrations and an appendix of the tool forms depicted in Diderot's *Encyclopedia*. This survey provides a guide to the hand tools and trades that played a key role in America's industrial renaissance. The iconography of American hand tools narrates the story of a cascading series of Industrial Revolutions that culminate in the Age of Information Technology.

The *Hand Tools in History* series is an ongoing project; new information, citations, and definitions are constantly being added as they are discovered or brought to the author's attention. These updates are posted weekly on the museum website and will appear in future editions. All volumes in the *Hand Tools in History* series are available as bound soft cover editions for sale at the Davistown Museum, Liberty Tool Co., local bookstores and museums, or by order from www.davistownmuseum.org/publications.html, Amazon.com, Amazon.co.uk, CreateSpace.com, Abebooks.com, and Albris.com.

Table of Contents

Introduction

Volume 13 in the *Hand Tools in History* series explores the iconography (imagery) of early American hand tools as they evolve into the Industrial Revolution's increased diversity of tool forms. The hand tools illustrated in this volume were selected from the Davistown Museum collection, most of which are cataloged in *An Archaeology of Tools* (Volume 9 in *Hand Tools in History*), and from those acquired and often sold by Liberty Tool Company and affiliated stores, collected during 40+ years of "tool picking." Also included are important tools from the private collections of Liberty Tool Company customers and Davistown Museum supporters. This volume is a guide to the hand tools and trades that played key roles in American history. They narrate the story of a cascading series of Industrial Revolutions that culminate in the Age of Information Technology. Following is a brief synopsis of the sections and their contents.

- Part I offers an overview of hand tools as simple machines and a description of the ferrous metallurgy of early American trades.
- Part II depicts the toolkits of early American trades, beginning with tools of forge-masters, multitasking blacksmiths, and related metalworkers. It also illustrates the Wooden Age's basic woodworking tools and a sampling of tools from early American trades, such as the cooper, farrier, wheelwright, sail-maker, cobbler, and others.
- Part III explores the iconography of representative hand tools manufactured in an industrial age that built machines that made some hand tools obsolete. It includes illustrations of the increasing variety of machine-made hand tools used to build, operate, or repair the tsunami of complex machinery that characterized the Industrial Revolution and American factory system it engendered. It concludes with illustrations of some of the most interesting machine-made tools and tools as sculptural objects.
- The text ends with an overview of the cascading series of Industrial Revolutions, the last of which is the Age of Information and Communication Technology. Observations are made about the continuing relevance of hand tools in the era of a hyper-digital global consumer society.
- The appendices include a representative selection of plates from Diderot's comprehensive encyclopedia of 18th century hand tools (Diderot [1751-65] 1964-6), illustrations of Roger Majorowicz's ax collection, and definitions and classifications of ferrous metals extracted from earlier volumes in the *Hand Tools in History* publication series.

After exploring the metallurgy and iconography of early American tools, *Tools Teach* reviews the American factory system of mass production of tools with interchangeable parts and includes an overview of the most important prime movers of the Industrial Revolution. The steam engine had multiple industrial applications, initially powering textile machinery, and then the railroad engine, and was responsible for the genesis of an industrial society that soon included a vast diversity of power tools, such as lathes, milling machines, drill presses, screw cutting lathes, planers, shapers, etc. The growing diversity of machinery in the American factory system led to the development of the automobile, an essential component of our domestic consumer society. The florescence of the global consumer society based on electronic equipment with no moving parts (radio, television, cell phones, computers, iPads, and the Internet) is, nonetheless, still dependent upon steam turbines (as in the steam turbines in nuclear power plants, aircraft engines, etc.) and other complex machinery as prime movers of industrial activity. Underlying this dependence is the necessity of the continued use of hand tools for maintaining the infrastructure of a complex, interconnected, globalized industrial society.

A familiarity with the hand tools of early trades and the history of technology provides an important basis for finesse in the use of hand tools for both the technicians of the age of information technology and participants in alternative sustainable economies. It is ironic that the evolution of a global consumer society driven by electronic technologies, digital communications networks, and nuclear-powered steam turbines continues to rely on many forms of simple machines, i.e. hand tools, to facilitate the ongoing operation of the infrastructure of modern society. Machinists, mechanics, electricians, plumbers, woodworkers, agricultural workers, and many other hand tool-wielding workers, craftspersons, and artisans are key to the viability of an increasingly complex industrial society. Expertise in using hand tools and awareness of their historical roots are vital components of economic and social stability. Artists, e.g. sculptors, metalsmiths, silversmiths, woodworkers, and printmakers, provide alternatives to the deadening boredom and psychic numbness of the electronic media of mass consumer society, the myriad benefits of electronic technologies notwithstanding.

10

Part I: Simple Machines and their Metallurgy

Simple Machines: An Overview

The function of all hand tools as instruments of manual operation lies in their efficiency as simple machines. Simple machines were first described by Greek philosophers, beginning with Archimedes' description of the lever, pulley, and screw. Later Greek philosophers added the wedge and windlass, the latter a form of the wheel and axle. During the Renaissance, the inclined plane was added to the list, defined as "the simplest mechanism that provides mechanical advantage… by changing the direction or magnitude of a force" (Wikipedia 2012). Contemporary paradigms of simple machines now include the lever, wheel and axle, pulley, inclined plane, wedge, and screw. The wedge is defined as two inclined planes positioned back to back and the screw as an inclined plane wrapped around a shaft or cylinder, explaining the function of common tools, e.g. screw augers, as simple machines. The growing variety of complex machines evolved during Renaissance were in fact combinations of simple machines working together.

During the Wooden Age, wind, water, and manual labor were the prime movers of an inefficient, often tedious, but sustainable pre-industrial craft-based society. An examination of Wooden Age toolkits reveals the overwhelming dominance of simple machines (see Figure 1 to Figure 6). The toolkits and chests of timber framers, shipwrights, joiners, and cabinetmakers reveal the overwhelming dominance of three simple machines: the wedge, screw, and lever. From axes, adzes, drawknives, pit saws, and planes to screw augers, calipers, and hammers, simple machines harvested and processed the natural resources that engendered the successful settlement of North America and the evolution of the Industrial Revolution. The lever (e. g. helve hammer of the finer), the pulley (e.g. block and tackle), and the windlass and anchor hoist (forms of the wheel and axle) were essential prime movers of a maritime culture that made exploration and settlement of the New World possible. The coasting trades that grew America in the centuries before the rise of industrial agriculture were an essential component of the successful birth of a new nation that would soon eclipse Britain in industrial production.

Complex machines such as printing presses, watches, and increasingly accurate navigation equipment were secondary components of an expanding mercantile society until the advent of the production of crucible steel and the widespread use of the steam engine. The appearance of crucible steel, so useful for the edge tools of the woodworker, ironically signaled the beginning of the end of the Wooden Age. Benjamin Huntsman's rediscovery of the lost art of making "cast steel" resulted in the production of a rapidly increasing variety of tool forms. By the mid-nineteenth century, advances in ferrous metallurgy technology and the evolution of the factory system of mass production created the hand tools illustrated in the third section of this text.

The figures below illustrate tools as simple machines that would have characterized the toolkits of artisans and workmen in 17th century colonial New England. Figure 1, the coach wrench, may not have appeared for use on the working wagons of maritime Boston until the 18th century. Malleable iron and steel in the form of simple machines were the prime movers of the wooden age (Figure 2, Figure 6, Figure 13 and Figure 14). One of a kind hand tools characterize the toolkits of early American industries (Figure 3, Figure 15, Figure 16). The design of wrenches would evolve from the wedge-based design of COACH WRENCH to the more modern screw-adjusted versions of the late 20th century (Figure 634). The ancient form of the turnscrew has many modern versions (Figure 9); have we forgotten that these modern implements are also simple machines? The evolution of the machine age resulted in the production of new forms of simple tools not seen before (Figure 7 and Figure 8). As simple machines evolved into complex machines driven by multiple simple machines (e.g. the car that drives Dasein) their basic function of providing a mechanical advantage—doing work—is often forgotten.

Simple machines form the core of implements used during recreational activities throughout history. The baseball bat (Figure 10) is a percussive tool of interest to many children and adults. Its roots derive from percussive instruments used by primitive hunter-gatherer societies (Figure 11). It is a simple machine whose accomplishments are sometimes memorialized in the annals of sports history. Simple machines in the form of ubiquitous hand tools (instruments of manual operation) continue to characterize our workshops and toolkits even in the age of information technology. The invisible ironmongers of sustainable economies and/or underground art festivals continue to swing their percussive hammers in obscure locations even as modern consumer society is entranced by ahistorical electronic media.

Figure 1. Coach wrench. Forged iron. DTM. 32103T4.

Figure 2. Socket chisel. Forged steel, malleable iron. Signed "J. BRIGGS". DTM. TCC2004.

Figure 3. Grafting froe. Reforged natural steel. DTM. 102409T2.

Figure 4. Wedge adjusted wrench. A simple machine that can be dated in its origin to the early Iron Age – 1200 BC. DTM. 21812T21.

Figure 5. Claw hammer. Bog iron. DTM. TAB1003.

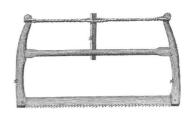

Figure 6. Buck saw. Illustration from Diderot: Carpenter. (Diderot [1751-65] 1964).

Figure 7. Surface gauge. Steel. Signed "V Oby." DTM. 21201T12.

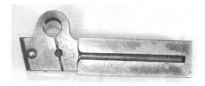

Figure 8. Lathe tool holder. Steel. Signed "WM AVERY & CO FOXBORO, MASS PAT. AUG. 11, 1908." DTM. 21201T12.

Figure 9. Screwdriver. Drop-forged steel, wood. Signed "Stanley Hurwood Pat April 01 Made in USA." DTM. 111900T11.

Figure 10. Baseball bat. Lathe turned ash wood. Signed "LOUISVILLE SLUGGER HILLERICH & BRADSBY CO. LOUISVILLE, KY. RUTH." DTM. 72013T1.

Figure 12. Lock and gunsmiths' metal file set. Steel. Signed "NICHOLSON." CLT. 82613LTC1.

Figure 11. Native American war club. Birch trunk and root ball. DTM. 21201T12.

Figure 13. Drawshave.
Illustration from Diderot:
Cooper. (Diderot [1751-65]
1964).

Figure 14. Compass plane.
Illustration from Diderot:
Carpenter. (Diderot [1751-65]
1965c).

Figure 15. Blueberry rake.
Sheet steel, wood. DTM.
72013T4.

Figure 16. Hub reamer auger.
Forged iron, wood. DTM.
72013T2.

The Identification of Hand Tools and their Use

The *Dictionary of American Hand Tools* (Sellens 1990) is an important reference used in the compilation of this text, providing a useful guide for tool identification. Sellens postulates the following classification of tool forms, which are utilized in our exploration of early American trades and the specialized professions of the Industrial Age that evolved out of them; our interpretation of the simple machines that executed these functions follows in parentheses.

- **boring tool**: any tool used to make a hole by rotary motion (screw)
- **striking tool**: any tool used to accomplish a forcing action by impact (lever, wedge)
- **cutting tool**: any tool with a sharp cutting edge (wedge)
- **measuring tool**: any tool used to gauge or scale a dimension (screw, lever)
- **marking tool**: any tool used to mark or scribe (wedge)
- **holding tool**: any tool used to hold a work piece temporarily in position (inclined plane, wedge, lever, screw)

- **guiding or testing tool**: any tool used to determine or describe a parameter (pulley, wheel and axle, lever)
- **sharpening tool**: any tool used to sharpen a cutting edge (wedge)
- **abrading tool**: any tool used to reduce a work piece by removing surface particles (wedge, wheel and axle)
- **supporting tool**: any tool used as a backup or support for a work piece (inclined plane, wedge)
- **spreading or piercing tool**: any tool used to widen a perforation or rift (wedge, lever)

The *Dictionary of American Hand Tools* (Sellens 1990) and Diderot's encyclopedia (Diderot [1751-65] 1964-6) are two among many important references cited in the *Hand Tools in History* publication series. Also among noteworthy sources of information about tools are R. A. Salaman's *Dictionary of Tools used in the Woodworking and Allied Trades, c. 1700-1975* (1975); Joseph Moxon's *Mechanick Exercises or the Doctrine of Handyworks* ([1703] 1989); Eric Sloane's many publications, including *Diary of an Early American Boy* (2004) and *A Museum of American Tools* (1964); Roger Smith's *Patented and Transitional Planes* (1981-92); and *The Chronicle* of the Early American Industry Association (1935-2012). The tool forms described in these and many other publications share a common, and now almost forgotten, ancestry of metallurgical progenitors, which gave birth to most modern tool forms.

The Ferrous Metallurgy of Early American Trades

Before depicting the many tool forms of early American industries, it is helpful to review the many furnace designs pre-dating Bessemer's revolutionary bulk process blast furnace. The variety of furnace and forge designs help narrate the story of the production of iron for hand tools from the simplest direct process bloomery furnace to the more complicated indirect process integrated ironworks typified by the Saugus Ironworks (1645). In all except the most primitive bowl and shaft furnaces, the furnace/forge juxtaposition characterizes almost all early ironworks. After smelting, the heterogeneous bloom of iron and slag from the direct process furnaces and the pig iron from the blast furnaces had to be reheated and reprocessed in nearby finery and chafery forges. It was in these forges that the iron bar stock for tool and nail production was refined and shaped before being sent to the multi-tasking blacksmith to make tools and hardware or to the nail-smith for that most important of all simple machines, the nail-as-wedge. These were the essential first steps in the production of hand tools that built a nation. A brief summary of furnace types helps illustrate the origins of iron hand tools and the evolution of an industrial society that perfected the art of the ironmonger in the classic period of American toolmaking.

Early Forges and Furnaces

Furnace types (before 1870): three basic types of furnaces characterize iron and steel production from the early Iron Age to the beginning of bulk process steel production (1870), the crucible, bowl, and shaft furnace (Tylecote 1987, Wertime 1962). Later Stucköfen and blast furnaces are larger forms of the shaft furnace, and cementation, reverbatory, and cupola furnaces are further adaptations of this form.

Crucible-shaped furnace: among the earliest furnace designs was a simple crucible, especially common in northern Asia and used for both cast iron and Wootz steel production. Multiple crucibles with as little as 1 kg capacity were used for Wootz steel production as early as 300 BCE. Cast iron production in larger crucibles in China can be dated at least as early as 800 BCE. The crucible-shaped furnace evolved into the cupola furnace commonly used in the 19th century to re-melt blast-furnace-derived cast iron. A late 18th century version of the crucible furnace consisted of two firebrick boxes used to manufacture cast steel for specialized purposes, such as watch springs and edge tools. The fire was in the lower box, and the upper box held the crucibles. Other forms of crucible furnaces had only a single firebox. Wasteful of heat and expensive to operate, crucible furnaces were made obsolete by the invention of the electric arc furnace (after 1900).

Bowl furnace: in the early Iron Age, the bowl furnace was the most widely utilized furnace for production of iron blooms from which iron bar stock was produced. The charge of ore was often located in back of the fuel, and reduction occurred via a current of carbon monoxide formed by the burning charcoal, assisted by a directed flow of air from a tuyère (the nozzle at the base of any furnace or forge for the admission of the air produced by a leather bellows or any other air-blast-producing device). The Catalan furnace of northeastern Spain is the most well-known form of the bowl furnace. Most earlier bowl furnaces were slag pit types. The rectangular open-hearth breakdown furnace used to smelt raw steel for Japanese swordsmiths is one of many variations of the bowl furnace form.

Shaft furnace: the shaft furnace with its many designs came to be the most commonly used furnace from the Roman era into the 19th century. Ore and fuel are mixed together in a four-sided shaft. The resultant bloom of wrought or malleable iron is extracted from a hole at the bottom of the furnace. Combustion is also aided by the use of the tuyère, powered by varying types of blowing devices.
Shaft furnace types are further subdivided as being slag pit types and slag-tapped types.

Slag pit furnace: furnace in which the slag accumulates *in situ* at the bottom of the furnace until the furnace is moved to an adjacent slag-free location.

Slag-tapped furnace: furnace in which the slag from the smelting process is withdrawn through a tap hole to an adjacent cavity or hollow in the ground. The slag-tapped furnace was more practical and efficient than the slag pit furnace.

Stucköfen furnace: low shaft (1 to 2 meters in height) slag-tapped furnace that characterized most Roman era ironworks. Low shaft furnaces gradually grew in height, capacity, and efficiency to become the high shaft Stucköfen of the German Renaissance, the immediate predecessor of the blast furnace.

Blast furnace: a form of high shaft furnace designed specifically to operate at high temperatures, thus carburizing iron ore to produce liquid cast iron. Small blast furnaces such as that used at the Saugus Ironworks were the essential component of indirect process "integrated ironworks," which also include fineries and chaferies for decarburizing the cast iron produced in the blast furnace. The wrought and malleable iron bar stock produced in the finery and chafery would then be made into tools in forges associated with indirect process ironworks, as typified by the Saugus facility. Small blast furnaces, such as those constructed in Carver, MA (> 1720), were also associated with onsite foundries that then cast stove and chimney parts, iron pots, and cast iron cannonballs. Blast furnaces gradually increased in size from the high shaft Stuckofen furnace to modern furnaces, typified by the Bessemer blast furnace, with a capacity for converting up to 50 tons of cast iron to low carbon steel.

Two variations of the shaft furnace, designed to prevent fuel-ore contact, the cementation and reverbatory furnaces, appeared during and after the 17th century. During the 18th century and until the 1870s, these two furnace types dominated blister steel and puddled, wrought, and malleable iron production until the appearance of modern bulk steel furnaces. Both facilitated greatly increased quality control and quantity of steel and iron production.

Cementation furnace: sandstone furnaces, in which iron bar stock was enclosed in a firebox before being carburized into blister steel; called "steel furnaces" in the United States in the Colonial period where they began appearing after 1720. First noted in Nuremburg in 1601, such furnaces protected the steel being smelted from the oxidizing influence of the burning fuel. The cementation furnace should not be confused with the many modern forms of steel furnaces developed after 1860.

Reverbatory furnace: a larger variation of the cementation furnace with a fire pit located underneath the hearth that facilitated the decarburization of large quantities of pig iron by heat reflected from the metal roof of the furnace, preventing fuel to ore contact. Primitive forms of the reverbatory (refractory) furnace, also known as the puddling furnace, were improved by Henry Cort into the modern form of the reverbatory furnace in 1785, allowing production of large quantities of high quality wrought iron.

Modern furnaces: during the period between 1830 and 1870, rapid industrial growth and the spread of railroads resulted in a greatly increased demand for iron and steel products. Blast and reverbatory furnaces were working at their maximum capacities. On August 13, 1856, Henry Bessemer proposed his new strategy for making malleable iron (low carbon steel) without fuel. His design of the modern blast furnace, which could convert 50 tons or more of cast iron to low carbon steel, was a key component in the rapid spread of industrial machinery and railroads throughout North America. The Bessemer process was soon supplemented by the Siemens-Martin and then the Siemens open hearth furnaces, the latter of which were particularly useful for a wide variety of iron and steel products, including tempered alloy steels, which fostered a virtual explosion of hand tool and machinery production after 1870. The essays on early modern metallurgy in *Steel- and Toolmaking Strategies and Techniques Before 1870* (Brack 2008b, 71-128) provide a detailed description of these innovations in iron and steel production. Also see *Handbook for Ironmongers: A Glossary of Ferrous Metallurgy Terms* (Brack 2013).

Cupola furnace: a nineteenth century form of blast furnace used for re-melting cast iron. It is a simple, fuel efficient shaft furnace usually using coke as fuel, one of the principal means for melting cast iron for foundries. The iron is melted in contact with the fuel. Only grey cast iron is produced in the cupola furnace. Instead of the stone pyramid form of the traditional blast furnace, cupola furnaces are constructed of sheet iron shells culminating in a narrow top with adjacent elevators and a bridge to bring fuel and ore to the top of the furnace.

Electric arc furnace: a furnace in which the heat needed to smelt metals is produced by an electric arc between carbon or graphite electrodes and the furnace charge; also called a low frequency induction furnace. It is one of the two most important 20th century steel-producing strategies, made possible by an electric power grid, the prime mover of which was biomass-fueled steam turbines or water-powered turbines. Its special advantage is that it avoids the oxidation of alloys, such as chromium or nickel, in the production of high grade tool steels. This furnace manufactured the great majority of the hundreds of varieties of useful high grade alloy tool steels produced in the 20th century. Its appearance was significant for edge toolmakers because the steel produced in the electric arc furnace gradually replaced the crucible steel production process after 1900. By 1940, the disappearance of both the puddled iron process and the crucible steel process seems to coincide with the diminishing quality of edge tools, at least until the recent perfection of cryogenic alloy steel edge tools. The electric arc furnace can be built in any size and utilizes scrap iron and steel efficiently. The first electric arc furnace was constructed by Sir Charles William Siemens in 1878 but was used for melting only. The first smelting of iron ore in the electric arc furnace occurred in 1898. In 1927, the first high frequency electric furnace was constructed. The uniformity and the high quality of alloy steels produced in electric arc furnaces is illustrated in the wide variety of sophisticated products of an atomic age (e.g. aircraft engine parts, nuclear power plant steam turbine blades, etc.) Since 1960, several basic oxygen process (BOP) furnace designs have supplanted, but not replaced, the multitude of electric arc furnaces.

The first three volumes of the Davistown Museum's *Hand Tools in History* publication series review the evolution of iron-, steel-, and toolmaking strategies and techniques utilized prior to the appearance of modern bulk steelmaking processes. Nonetheless, a brief synopsis of earlier iron-making technologies helps illuminate how the metal hand tools of the Wooden Age came to be fashioned. The history of iron production can be summarized as a description of two processes, both of which were used to produce the iron bar stock that blacksmiths and specialized toolmakers would use to forge the simple machines that were the key element of the successful settlement of the New World, the direct and indirect processes of iron-making.

Iron-Making Techniques

The direct process of iron-making has roots in the height of the Bronze Age (1900 BCE) when the Chalybes made steel edge tools in northern Turkey on the south shore of the Black Sea, 700 years before the sudden advent of the Iron Age. (See the chapter on the Evolution of Iron Metallurgy in *Steel- and Toolmaking Strategies and Techniques before 1870* (Brack 2008).) Ancient iron smelting sites (furnaces) can be identified at archaeological sites in many locations in Europe, North America, the Near East, and Africa. The blooms smelted in primitive early furnaces, retrieved from the furnace bottoms as "loupes" with some slag eliminated by the helve hammer, ranged in weight from 5-15 kg in the smallest bloomeries to 50-100 kg for the larger currency bars transported down Europe's famous Iron Road. These blooms provided the raw materials that would be reworked by the forge-master to produce the iron bar stock utilized by blacksmiths to make tools. Usually consisting of a heterogeneous mixture of low and medium carbon iron, slag (ferrous silicate usually dominates the slag residue,) and globs of steel, the forge-master forged the iron loupes from primitive inefficient smelting furnaces into iron bar stock and sheet iron. These raw materials were the universal components of the simple machines made by metal toolmakers throughout the Iron Age. From the products of the forge-master, the blacksmith forge-welded simple machines, including crowbars, hammers, tongs, calipers, axes, adzes, chisels, and drawknives, as well as brush hooks, spades, hoes, eel spears, and knives. In early primitive ironworks a multitasking ironmonger could be the bloomsmith, forge-master, and toolmaker. In many cases, blooms of iron, steel, and slag were often transported long distances, sometimes as ballast in ships, to provide the raw materials for the forge-master, who sometimes shared the same hearth and bellows with the toolmaker, if he, himself, was also not a toolmaker.

The Frobisher expedition to the Baffin Islands in 1576 provides a well-documented example of the transatlantic movement of blooms from distant, if unknown, bloom smelting furnaces to one of the first forges in the New World. The recent excavations at the Fort Popham Colony in Maine clearly document the typical bloom smelting furnace-forge juxtaposition, where local bog iron was smelted into blooms, leaving a telltale residue of silicate slag, and then processed into iron bar stock, shipsmiths' fittings such as nails, spikes, and possibly some tools at a nearby forge (Brain 2010). Forge residues differ from the silicate slag of the bloomery furnace, with fragments of iron bar stock and tools identifying such a site as independent of, even if adjacent to, the bloomery furnace that smelted the forge-masters' raw materials. During the period between 1650 and 1840, the bog iron country of eastern southern New England was the location of hundreds of bloomery furnaces, as well as numerous small blast furnaces, which then supplied both adjacent and distant fineries, chaferies, toolmaking forges, and cupola furnaces with the blooms and cast iron billets for their products. If not made into tools on the spot, the bar stock of the forge master could be traded to farmers for food or sold for cash. Many a New England farmer was equipped with a small forge with accompanying anvil and post vise for toolmaking, as evidenced by the ubiquitous appearance of these tools even in the first decade of the 21st century (Figure 29). Before the Connecticut and Pennsylvania rock iron ore industries began their vast production (after 1725,) iron bar stock from New England's bog iron bloomeries and forges was a common trade item on the coasting vessels that made Boston North America's busiest port ($\pm$1700). This coasting trade continued until the early 19th century, supplying a decreasing percentage of iron bar stock used in America after 1725.

The Indirect Process of Iron-Making

Bloom smelting was a convenient but inefficient method of producing the raw materials needed by the forge-master. A few blooms a day (3-5), initially with a weight of $\pm$5 kg, were not sufficient to supply the metallurgical needs of New England's rapidly growing population, nor the thriving market for iron bar stock that quickly evolved in the colonies south of New England. While blooms and currency bars smelted in Europe had been imported to the West Indies in the 16th century and to the American colonies in the 17th century to supplement imported tools, the rich bog iron resources of southern New England gradually supplanted the importation of English iron bar stock after 1665. Nonetheless, American toolmakers continued to rely on the highest quality Swedish and

English malleable iron and German steel, and, after 1700, English blister steel for toolmaking purposes.

The well-documented importation of iron and steel from England and Europe in the two centuries following the establishment of the Saugus Ironworks has tended to obscure the important role the bog iron deposits of southeastern Massachusetts played in the rapid evolution of an indigenous iron industry. The robust Colonial exploitation of these deposits was an important component of the initial success of a shipbuilding industry that led, in turn, to a vigorous coasting trade where lumber, codfish, and agricultural products were keys to the success of a growing market economy. The presence in New England of multiple indirect-process integrated ironworks, as well as direct-process bloomeries, was an important component of the early Colonial history of New England.

After settlement of the Plimouth Plantation and the great migration of settlers to Boston and eastern New England between 1629 and 1643, the indirect process of iron-making quickly made its appearance in Saugus, MA, in 1645. The blast furnace built there by John Winthrop was among the most modern in the world. Staffed by experienced English ironmongers, these integrated ironworks were characterized by the following components.

- The **blast furnace**, including that at the Saugus Ironworks (Hammersmith, MA), burned its charcoal fuel more efficiently than bloomery furnaces, producing higher temperatures that enhanced carbon uptake by the iron ore. The production of liquid cast iron, which has a lower melting temperature than wrought or malleable iron, was the result. Liquid slag, which was drawn off the top of the liquid cast iron, also characterizes blast furnace operation. At Saugus and all other blast furnaces, the liquid slag was removed as waste, and the melted iron, now high in carbon, was directed to shafts and molds, where billets and pigs of cast iron were created.
- An adjacent **finery**, which could reheat the billets and pigs to various temperatures, decarburized the cast iron by removing carbon to create the malleable iron (>0.08 % carbon content [cc] to <0.5 cc) or wrought iron (<0.08 cc) needed for specific iron-making applications.
- **Helve or trip hammers**, omnipresent at all furnaces, were used to hammer out the slag impurities as part of the fining process.
- An adjacent **chafery** would reheat and further refine the iron, allowing it to be shaped or rolled into specific forms, such as sheet iron for nail-making or bar stock for toolmaking.
- **Toolmakers**, such as Joseph Jenks at Saugus, worked at an adjacent forge at the larger ironworks, forge welding such items as axes with carburized cutting edges or ductile wrought iron eel spears and hay thieves, always for a Colonial economy

that had a rapidly growing need for tools for shipbuilding, sailing (e.g. marlin spikes), fishing, or agriculture.

John Winthrop constructed another blast furnace at Quincy, MA, in 1646. Restored in the 1940s (see photo in *Art of the Edge Tool*; Brack 2008a, 19), this historic site now underlies a golf course, which was bulldozed above it in the 1960s, just north of the Furnace Brook Parkway. Other integrated blast furnace complexes were constructed in the decades after 1645 at Taunton, Pembroke, Hanover, Dartmouth, Bourne, and elsewhere, and later in Carver, Middleboro, and many other locations. A detailed preliminary survey of the foundries and forges in the bog iron country of southeastern New England is listed in Appendix 2 in *Art of the Edge Tool* (Brack 2008a).

All southeastern New England furnaces and forges had the advantage of vast deposits of nearby bog iron, in contrast to the Saugus Ironworks, which, having quickly depleted nearby bog iron deposits, eventually failed due to the expense of shipping bog iron from the North River, the Jones River, and especially the Taunton River in southeastern New England. The cranberry bogs of Plymouth County are a testament to the vigor and extent of New England's Colonial iron industry, which was soon supplanted and then replaced by the vast rock iron deposits in Connecticut, Pennsylvania, Maryland, New York, and other nearby states. A surprisingly large number of hand tools used in the century after 1660 to build our nation were forged in the colonies, especially at New England's many water-powered forges, and not imported from England.

The Puzzle of Natural Steel

Figure 17. Framing chisel. Forged iron and weld steel. DTM. 42904T3.

One of the curiosities of the historical interpretation of our past is the romantic idea that early ironworkers (the bloomsmith is not often differentiated from the forge-master) produced only pure wrought iron (<0.08 cc) from their labors. In reality, the iron blooms produced in small quantities but at many locations were a mixture of wrought iron, malleable iron (>0.08 cc), slag, and occasional pockets of natural steel (>0.5 cc). A knowledgeable forge-master could, depending on the needs of his customers, have used the now forgotten tricks of his trade to produce from his bloom wrought iron, low-carbon malleable iron, and higher carbon malleable iron. Low-carbon wrought iron was often sought by blacksmiths for making hinges, calipers, cheese whips, eel spears, and other horticultural tools where the ease of forging ductile twisted or decorated objects was more important than the strength and hardness needed in the higher carbon malleable iron tools, such as rock bars, shovels, hay forks, and blacksmith hammers, which they also produced. The ductility of low-carbon wrought iron could quickly render a hardworking horticultural tool useless. Needing even a higher carbon content (>0.5% cc) were the steel edge tools so essential for timber harvesting and shipbuilding. A surprising number of Colonial era edge tools were made from the natural steel produced by carburizing the direct process bloom.

Their relatively high slag content and primitive forge-welded appearance make these tools obviously different from the more sophisticated edge tools imported from England or made with German steel imported from England and Europe. After 1700, blister steel supplanted the use of German steel by English edge toolmakers.

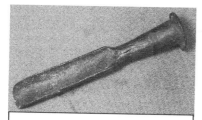

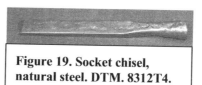

Figure 19. Socket chisel, natural steel. DTM. 8312T4.

Figure 18. Forge-welded gouge. Iron and natural steel. A product of a working shipsmith. DTM. 10407T5.

Figure 20. Complex molding plane. Wood, forged steel blade. Robert Wooding. DTM. 50402T1.

A conundrum is also present with respect to Colonial edge tools made after 1700. Both steel bar stock and "steeled" edge tools were very expensive to import from English and Continental sources. The knowledgeable ironmongers who came from England in the great migration were as sophisticated as any in England and Germany. Trained in the tradition of making (German) steel from partially decarburized cast iron, these toolmakers could certainly make steel by decarburizing and fining New England-made cast iron into steel as the need arose after 1650. They could also take the direct process natural steel of the bloomsmith, more difficult to refine, and tediously reforge it into small quantities of relatively high quality steel for edge tool production. We know from our urban and coastal archaeological excavations that the vast majority of the larger edge tools used by New England's shipwrights, especially broad axes, side axes, mast axes, felling axes, mast shaves, drawknives, and to a lesser extent adzes (those with English marks, e.g. James Cam, are more common than other larger edge tools with English marks) were forged in New England, especially after 1700. The puzzle remains: to what extent early edge toolmakers were able to refine and reforge readily available and much cheaper natural steel from the direct process bloom? In this context, an elite class of edge toolmakers, such as the Saugus toolsmith Joseph Jenks, supplemented the output of more primitive New England toolmakers, who were often local blacksmiths and shipsmiths. These elite forge-masters and edge toolmakers had the ability to forge high quality steel from the natural steel of the bloom and to carburize malleable iron, then harden, temper, and anneal this steel during the forge-welding of edge tools. Many edge tools made from direct process-derived natural steel cannot now be differentiated from tools made from domestically-produced or imported German steel (made from partially decarburized cast iron), nor, after 1710, from the growing imports of cementation (blister steel) made by the new process of carburizing malleable iron in the cementation furnace. Much larger quantities of edge tool steel could be made from these processes. But small quantities of natural steel from the bloomery furnace, because it was so readily available, were produced in New England and elsewhere and refined in small forges and ironworks. The last direct process ironworks ceased operation in isolated areas of Appalachia, where making the Kentucky rifle by hand lingered into the first decades of the 20th century.

The Ironworks at Taunton, Pembroke, Middleboro, Litchfield, etc.

Soon after Colonial America first integrated ironworks was established on the Saugus River (1645), other ironworks followed at Braintree, Quincy (Furnace Brook), Pembroke, Taunton, Middleboro, and numerous other eastern Massachusetts locations. Among the most famous of the eastern New England ironworks was the Leonard furnace and forge complex on a tributary of the Taunton River in what is now Raynham, MA. Historical sources describing these ironworks, including town or local histories, often do not differentiate between the direct process bloomery furnace and associated forges, and the more complex integrated ironworks. The furnace at Braintree (1645), which included an adjacent finery, was a smaller direct process ironworks and typified the many furnace-forge examples that were established at water sources throughout southeastern New England in the next century. The Furnace Brook example in Quincy (1646), was probably an integrated ironworks. The Leonard ironworks in Taunton appears to have started as a bloomery before growing into an integrated ironworks, as did the first furnace established at Middleboro in 1692. This ironworks was enlarged into an integrated ironworks, eventually owned by the English Loyalist, Peter Oliver (Weston1906).

Early Colonial ironworks were not limited to the bog iron country of southeastern New England. One of the first documented used of rock iron ore was at Cumberland, RI, 1725. By the 1730s, as illustrated by the complexity of the various ironworks in the Litchfield, CT, area, which were now using local rock ores to supplement, and then replace, bog iron, blacksmiths as well as nail-makers and anchor forge-masters were working at or adjacent to ironworks that produced rolled malleable sheet iron for nail-making in finery

24

associated rolling mills. At the same time, founders at the Litchfield Ironworks were clearly smelting cast iron for tool and equipment production. Molten iron (cast iron) was "drawn into draughts" (molds), which constituted the key step in casting plows for the farmer, or the "sleigh shoes," the runners for a sleigh (Warren 1992, 11). The ironworks at Litchfield, which flourished for over a century after the establishment of numerous ironworks near the bog iron deposits of eastern New England, included the simultaneous production of both cast iron and malleable iron, suggesting either that the bloomery furnace also produced cast iron or that small blast furnaces were also part of these early ironworks. Commenting on the Litchfield area Salisbury ironmasters, Gordon (1996) notes:

> By 1770, Samuel Forbes in East Canaan was selling forge hammers, gun barrels, standardized parts for gristmills, and anchors through agents located in the principal colonial cities. With his partner John Adam, he built a rolling mill in 1780 and another in 1794 and added nails and saws for nearby marble quarries to the product line. By the end of the 18th century, ironmasters in Litchfield county alone had 50 forges and three slitting mills in addition to their blast furnaces; there were more in Massachusetts and New York (Gordon 1996, 65).

In response to New England's rapidly expanded ironworks, as well as those to its south in Pennsylvania, Maryland, New Jersey, etc., the English Parliament passed the *Iron Act of 1750*, demanding that the colonists "abate any and all mills, engines, forges, or furnaces" (Warren 1992, 8). As a result, written documentation about the location, design, and functioning of early American ironworks before 1785 is rare or nonexistent. The tool forms produced at American furnaces, forges, and foundries linger long after the bloomsmiths, forge-masters, and toolmakers, as well as the owners and users of these tools, have passed on, providing clues to the iron- and steelmaking strategies and techniques that lead to their creation.

Imported Hand Tools

Many tools used by early American artisans and still found in old workshops and collections were brought to America by immigrants from Colonial times to the early 20th century, as for example, the English planemaker Robert Wooding's complex molding plane (Figure 20). As with American-made hand tools, all tools made before 1840 were handmade, and if made of iron or steel, usually forge-welded. These implements, often one of a kind, were the essential instruments of manual operation for settlers in the new-found-lands. Particularly commonplace were the many hand tool forms made in English toolmaking centers such as Birmingham, and later, Sheffield. Such tools are usually stamped with English touchmarks. Figure 21, illustrations from Moxon's *Mechanick Exercises or the Doctrine of Handiworks* (1703 [1989]), depicts commonplace tool forms still frequently encountered in contemporary tool collections. The top two joiner planes with the loop handles are now very rare; the ax forms are also seldom encountered. For the next three centuries, the remaining tools and work bench retain these basic forms.

A Directory of Sheffield (Rees [1787] 2004) provides an interesting listing of hundreds of toolmakers working in Sheffield prior to 1787, including cutlers, scissor and file-smiths, edge toolmakers, and sickle makers. Maker's signatures and typical touchmarks are included, the latter of which clearly differentiate English-made tools from those produced by American toolmakers. Also occasionally recovered from old New England workshops are files, phlemes, pen and pocket knives (the text has a huge listing of these artisans), razors, saws (Figure 27 and Figure 34), scissors, scythes, sickles, shears, and table knives. Plane blades and carving tools are among the most commonly encountered English tools. Among the most prolific of English toolmakers was Peter S. Stubs (working dates 1777-1806), whose hand vises (Figure 24), files, and other tools make frequent appearances in American tool chests (Figure 103, Figure 104).

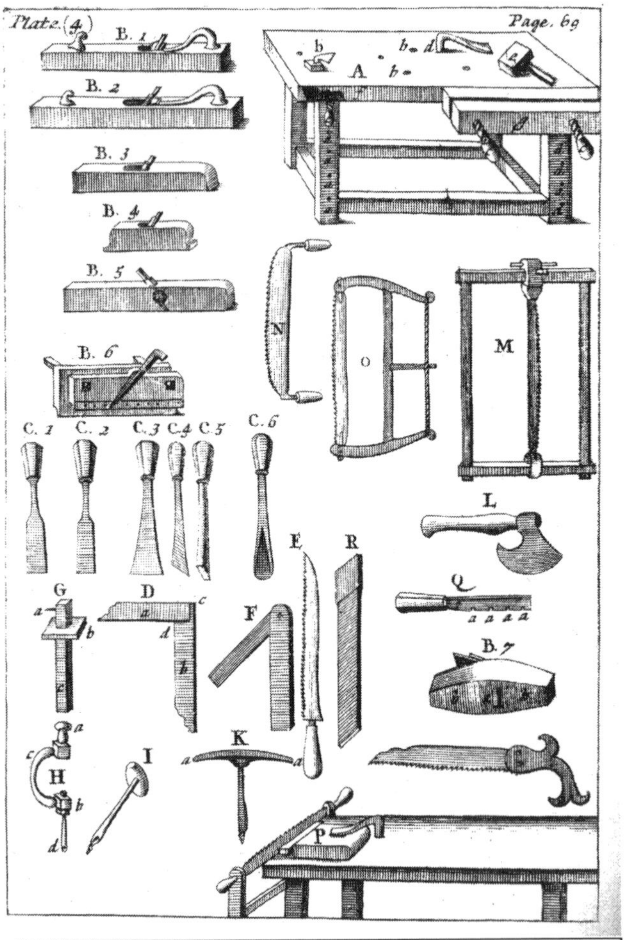

Many English tool forms were derived from European sources, including Germany, Italy, Spain, and especially France. American-made farriers' buttresses are a commonly encountered, if obsolete tool. The one illustrated (Figure 23) is a distinctively European form, and was probably brought to America by immigrants from Germany. The hand adz (Figure 22), a common tool form in southern Europe, is notable in that this particular form was not copied by English or American toolmakers despite its common presence in the toolkits of Spanish and Portuguese woodworkers. Renaissance France is a most important source of tools, which were first copied by English toolmakers before making their appearances as products of indigenous American toolmakers. Once drop-forging technology was used to make tools in Europe, many a signed tool made in

Figure 21. (Moxon [1703] 1989, 69). This illustration shows the typical tools and their designs in the 17th century.

26

France, Germany, and Italy were brought to the United States (Figure 400). Some of these are illustrated in *Part III*.

Figure 22. Hand adz. Forged steel, iron, and wood. DTM. 41505T1.

Figure 23. Farriers' hoof paring tool. Courtesy of Douglas Schlicher.

Figure 24. Hand vise, iron and forged steel, signed "P Stubs". Iron and forged steel. Peter S. Stubs. DTM. 62406T7.

Figure 25. Combination calipers. Cast steel or German steel with brass knob and copper rivet. Peter S. Stubs. DTM. 22411T24.

Figure 26. Door handle. Wrought iron. DTM. 072112T7.

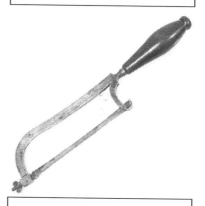

Figure 27. Jewelers' hacksaw. Forged steel, wood (cocobolo), brass. CLT. 31713LTC1.

Part II: The Toolkits of Early American Trades

Diderot's Encyclopedia

Diderot's *Pictorial Encyclopedia of Trades and Industry* ([1751-65] 1964-66) is one of the primary sources of illustrations depicting early tools. Other important early sources pertaining to the iconography of early American hand tool forms include Moxon's *Mechanick exercises or the doctrine of handiworks*(1703) and the English pattern books, including Smith's *Key to the various manufactories of Sheffield* ([1816] 1975). With its eleven volumes of plates, Diderot's *Encyclopedia*, provides a particularly comprehensive survey of images of European-made tools, which characterized the craft and industries of the Enlightenment in France and the rise of scientific rationalism that was, along with English empiricism, the philosophical basis of the Industrial Revolution. A selection of the most interesting plates from Diderot is reprinted in *Appendix 1*. Many of the tool forms illustrated in Diderot were imported to America, sometimes after being reproduced by English toolmakers, but were also soon produced by American toolmakers. Diderot includes images of the equipment of many trades of the late 17th century and first half of the 18th century. Some were of ancillary interest to the successful Colonial effort to displace the once vigorous Native American communities who preceded European settlement (e.g. luthier, dyer, baker, barber, scale-makers, clothier, etc.) Other trades listed in Diderot depict many of the tool forms used in early American industries, illustrating the links between the commercial and social labyrinths of a flourishing Renaissance mercantile monarchy, France, and the rapidly expanding communities of the North American continent. England was an intermediary in this linkage, adapting or copying the tool forms illustrated in Diderot, many of which soon reappeared in Colonial America. England was France's only competitor for exploration and settlement of North America. After the fall of Quebec in 1759 and the English victory over the French for control of North America, much changed, except the toolkits of Europeans settling in North America.

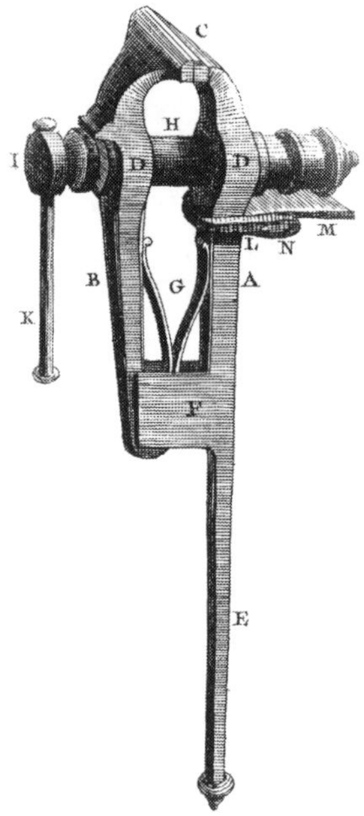

Figure 28. Post vise from Volume 2, "Swordsmith," in Diderot ([1751-65] 1965a).

One of the most ubiquitous images in Diderot is the post vise (Figure 28). Illustrated as an essential component of the trades of carpentry, cutler, spur-smith, swordsmith, saddler, and metal toolsmith, post vises were soon forged in North America in a surprisingly uniform design. Initially forge-welded, the post vise had cast and/or malleable iron components, as well as wrought (puddled) steel jaws, the manufacture of which predated modern Bessemer steel production and open hearth steel production techniques (1870) by a half century. The steely jaws of the ubiquitous American post vises were probably made either with domestically-produced German steel (decarburized cast iron) or puddled steel from refractory furnaces after 1820, rather than from more expensive blister steel made in the US since 1720. The records of the Litchfield Ironworks document the domestic production of cementation steel as well-established in Colonial America in the 18th century. A 200 weight quantity is noted as a delivery item along with locally made nail rods to the Milton, CT, ironmonger Lot Chase in 1789 (Warren 1992, 15). After 1875, steel made in Siemens open hearth furnaces replaced the steel produced by the earlier diversity of, and now forgotten, steelmaking techniques used for most tool production. The one exception to this observation is the continued reliance on cast steel for edge tool production until the electric arc furnace made cast steel obsolete (after 1925).

The Forge-Master and his Progeny

An examination of the toolkits of early American trades must begin with the blacksmith and the forge-master, who enabled the blacksmiths' multitasking. The first blacksmiths as toolmakers were also forge-masters who fined the bloom from the smelting furnace into iron bar stock for their own toolmaking. This tradition characterizes toolmaking at the beginning of the Iron Age (1200 BCE) and lingers in Colonial New England and elsewhere at small isolated ironworks despite the presence of integrated ironworks such as those at Hammersmith (Saugus, MA) and Middleboro (Oliver Forge).

Whether working at the early bloomery furnaces of Europe or in America at New England's many direct process furnace-forge ironworks, all bloomsmiths faced a fundamental challenge inherent in the production of a bloom of any size. First and foremost, a bloomsmith had to maintain careful control of his furnace gasses. In *American Iron*, Gordon notes the challenges faced by the bloomery smelter:

> To attain the high temperature necessary to melt the slag the bloomer had to have a strong flow of air through the fuel. This increased the amount of CO^2 in the furnace and could lower the ratio of CO/CO^2 below what was needed for smelting. A bloomer had to achieve just the right balance between air flow and heat loss to get simultaneously the necessary temperature and CO/CO^2 ratio. Additionally, he had to have rich ore because he got metal only from the iron oxide that remained after the slag formed. Bloomery slag had to contain about 55% iron to be liquid in the hearth: if the ore had less than about 55% iron… no metal would form (Gordon 1996, 91-2).

In Colonial New England, many a forge that would be used to re-melt and further refine the bloom of raw iron from the bloomery smelter was located at the site of the smelting furnaces. In some cases, only one or two ironmongers operated the simple furnace-forge complexes in southeastern New England's bog iron countryside.

At some point in the early Iron Age, the furnaces that smelted the iron blooms, and, later, cast iron billets, were not necessarily at the same location as the fineries and toolmaking forges that produced the final products of the ironmonger. Traditionally, most multitasking blacksmiths utilized iron stock in the form of rods, bars, or sheet iron to make their tools, guns, or nails. The direct process bloomery smelting furnace produced loupes of iron that had to be processed at a forge into these useful shapes.

The forge-master used essential simple machines to forge the bloom from smelting furnaces into the bar stock and sheet iron for toolmakers: the hammer (either helve or trip), bellows (also a lever), anvil as supporting tool, and water wheel as the prime mover of forge operations. Every forge needed bellows, whether hand operated or powered by water wheels. Eric Sloane provides a graphic illustration of the four varieties of water wheels used in early American mills and forges (Figure 29). Every forge also needed helve or trip hammers

Figure 29. Mill wheels, from *Diary of an Early American Boy, Noah Blake 1805*. Eric Sloane (2004) Dover Publications.

to help remove the silicate slag and other contaminants from the bloom and consolidate the iron into a more uniform consistency. Multitasking blacksmiths utilized anvils, hammers, and tongs, the latter two levers as simple machines, as well as the cutting and sharpening tools noted below, to produce the edge tools that harvested the forest and built the ships and wharves of the Wooden Age.

Tylecote's (1987) illustration from Pleiner (Figure 30) provides a handy guide to six knife and edge tool forms still encountered in New England in tool chests and collections. The simultaneous production of all edge tool forms described by Pleiner (1962) before the beginning of the early Industrial Revolution and its mass production of cast iron also characterizes all later efforts at edge tool production. Even the lowly all-iron ax produced by isolated smithies as recently as the mid-19th century continues to make its occasional appearance in New England collections. More commonly encountered is the simultaneous presence of the all steel (German steel – continental method) and steeled (weld steel) trade axes in tool collections and archaeological sites in eastern North America. The latter form became more common in the 18th century, probably in response to the sudden availability of blister steel bar stock made in cementation steel-producing furnaces, the use of which became widespread in England in the very late 17th century. Also still commonly encountered as the legacy of the multitasking blacksmith, including those with small shops in their barns or in the back streets of many towns, are the multiplicity of horticultural tools made from reforged steel rasps. Figure 33 is an example of a commonplace simple machine, the wedge, needed in every workshop. Also typical of the recycling of dull farriers' rasps are Figure 138 and Figure 139.

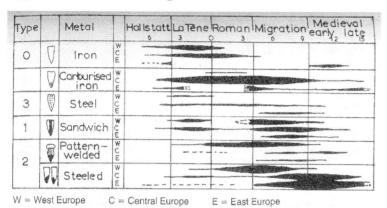

Figure 7.21 Chronology of knife forms (after Pleiner 1962).

Figure 30. Knife forms. From Tylecote (1987, 269).

Figure 31. Unfinished ax head with forge-welded edge. Forged iron and steel. DTM. 81411T1.

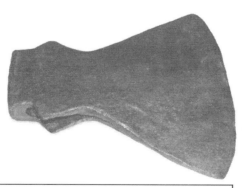

Figure 32. Broad ax. Forged iron. DTM. 111001T29.

Figure 33. Carpenters' wedge. Reforged steel. DTM. 21812T6.

The **blacksmith** as ironmonger soon evolved into specialized trades:

Figure 34. Keyhole saw. Steel, wood, brass. William Cresson. DTM. 81713T17.

- The **gunsmith** used anvil, hammer, mandrel, and horn to make that essential implement for frontier conquest, the Kentucky rifle, which displaced the matchlock in the mid-18th century. All marksmen relied on the lever as the primary source of leverage needed to activate the cocking mechanism and fire the gun.

- The forgotten **shipsmith** made edge tools before the forging of edge tools became a specialized trade at the end of the late 18th century. The simple machines he used were anvils, hammers, and specialized cutting and shaping tools, essentially the same as those used by the blacksmith as toolmaker.

- **Swordsmiths** and **cutlers** were forge welders of another genre of edge tools, a term that historically references woodworking edge toolmakers (see the trade listings in 19th century Maine business registries), who created the same simple machines as those used by the timber harvester and shipwright, but for their own specialized military and culinary missions and with their own unique metallurgical recipes. As noted by Pleiner (1962), the traditions of the pattern welding of knives, especially swords, such as those made by Damascus and Japanese swordmakers, have roots in the early Iron Age and flourished during the Roman period.

- The **nail-maker** used specialized slitting tools (levers) to cut sheet iron for nail-making. Anchor-making was a specialized ironmongering trade. The tools of the **anchor-maker** were the same simple machines used by all ironmongers though possibly larger in size; i.e. forge with bellows, water wheel to power the bellows and trip hammer, hammers, and anvils, supplemented by specialized hoists to lift the anchors.

- Only the **farrier**, maker of horseshoes and related hand tools, and the last of the ironmongers to hold forth in New England's towns and villages, is still remembered as a footnote to the appearance of that most complex of machines, the automobile.

Among the most important artisans of the Wooden Age is the ubiquitous **cooper**, whose strange edge tools still make their appearance at New England's flea markets. The cooper was an important facilitator of early American industries, making the staves for wine casks, salt boxes for the coasting trades, water and brine slack tubs used by the blacksmith for quenching, and an endless array of food storage containers for domestic and agricultural purposes.

More modern edge toolmakers, such as the Underhills, Buck Brothers, and Thomas Witherby, inhabit a later chapter in the story of the evolution of simple machines into the complex machinery that drove industrial society into the age of chemical fallout. These famous edge toolmakers forged both the exquisite tools still used in the fading years of the Wooden Age and the patternmakers' edge tools essential for the creation of the wooden patterns used to cast the machinery of the Industrial Revolution. The unique florescence of these edge toolmakers combined ancient traditions of forge welding and heat treatments with the use of modern (>1840) drop-forging machinery in an era when cast steel had not yet been replaced by the tempered alloy steels that began characterizing tool production after 1870. Crucible-derived cast steel continued to be used in edge tool production for another 50 years, even as Bessemer's blast furnaces began making the modern version of "cast steel," as in the structural steels of skyscrapers, steam engines, and the safer steel rails that replaced wrought iron rails after the Civil War.

The Toolkits of Multitasking Blacksmiths

The basic toolkit of the multitasking blacksmith included anvils, hammers, mandrels, tongs, cutters, shaping tools, such as fullers, flatters, swages, and forks, and files and rasps along with forge, bellows, and slack tubs for brine and water for hardening and tempering (see illustrations below).

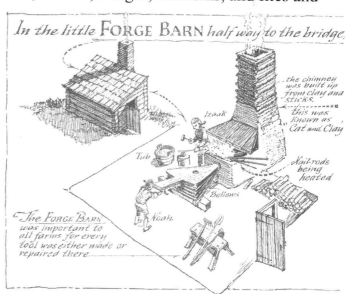

Documentation of the toolkits of the multitasking blacksmith thus must acknowledge the complete chain of sequence for the production of such tools. Before a blacksmith could produce any tools, the finery forge would shape and often carburize the raw iron products of the smelting furnace prior to tool and hardware production. The furnace used by a working blacksmith or shipsmith was much smaller than the smelting furnaces used in either direct process

Figure 35. Forge barn, from *Diary of an Early American Boy, Noah Blake 1805*. Eric Sloane (2004) Dover Publications.

or indirect process iron production and usually smaller than the finery forges, which processed the raw iron into the rods, bars, and sheets used by the blacksmith. A multitasking blacksmith working as a shipsmith, as at Fort Popham, Popham, ME (c. 1607), first reheated the bloom and shaped it into useful iron stock before fashioning it into the spikes, ring bolts, and hand tools needed to build small coasting vessels for exploration, repair existing pinnaces, and make hand tools.

The common mandrel is among the most important tools frequently used by a blacksmith (Figure 43 and Figure 44) for sizing metal products ranging from gun barrels to the ring bolts of the shipsmith and the iron hoops of the coopered cask. The anvil illustrated in Figure 37 has an excellent horn, a form that serves many but not all of the functions of the mandrel. Tongs of various designs were used to manipulate bar stock on the anvil during the forging of tools and hardware (Figure 57, Figure 58). Hardies and other cutting tools were used to cut the bar stock being utilized for toolmaking (Figure 59 to Figure 64). Shaping tools, such as flatters and swages (unique forms of the hammer), were used for shaping (Figure 67 to Figure 72). Files and rasps were used for the final finishing touches of the toolmaking process (Figure 73 and Figure 74). Bealer's *The Art of Blacksmithing* (1976) is among the most straightforward and easy to use references explaining how the blacksmith used these tools to fashion objects essential for the survival of farmers, for example, who could not afford to import expensive agricultural implements, which were easily fashioned by the local toolmaker. In many cases, the forge and workshop of the farmer was the location of tool making and tool repair, as illustrated by Eric Sloane's depiction of a farm forge (Figure 35).

Anvils and Swage Blocks

The anvil was the most important of all tools used by the multitasking blacksmith. The history of anvil-making begins with blooms of roughly shaped iron, usually formed into rectangular or square masses. All metalsmiths as multitasking blacksmiths needed an anvil as the fundamental component of their toolkits. This includes the edge toolmaker, gunsmith, silversmith, nail-maker, whitesmith, and farrier. This simple machine, which Sellens (1990) defines as a "supporting" tool, was the fundamental component of all metalworking before the appearance of drop-forging machinery gradually replaced many of its functions after 1840.

In the early Iron Age, and until about 1600, the anvil was often a squared lump of either high carbon malleable iron or cast iron. Beginning in the Renaissance, steel plates began to be welded onto the surface of these primitive anvils, making them much more durable than an iron anvil. No records currently exist to tell us when this event first occurred. At some point in the early Renaissance, horns were added to anvils, probably for the shaping of gun barrels. By the mid-18th century, English anvils of a "modern" design were being produced and imported to North America (Figure 37). According to Postman (1998), all anvils were imported to America before Fisher & Norris began making anvils in New Jersey in 1847. This is a dubious assertion as Colonial ironmongers had the materials and knowledge to fashion primitive wrought or malleable iron anvils that are not distinguishable from those imported from England. Once Fisher & Norris began utilizing factory system technology to make anvils, the predominant form of anvil described in Postman is the "steel-faced wrought iron anvil" (Postman 1998, 30). Postman notes that:

- cast iron anvils were also manufactured and had been in use since the 14th century, coinciding with the introduction of the blast furnace in Europe. Fisher & Norris made some cast iron steel-faced anvils, which are considered second quality anvils and lack the ring and bounce of the higher quality wrought iron and cast steel anvils.
- cast steel anvils were produced in Sweden "around the turn of the 20th century" and "were made of high-carbon open hearth steel and had a hardened face."
- "As wrought iron became more expensive than the low carbon steel in the 1930s, one American anvil-maker (see *Arm and Hammer*) used a low carbon cast steel base, a forged low carbon steel body, and a high carbon tool steel plate." (Postman 1998, 33).

Despite the many variations in the ferrous constituency of anvils, and the many obscure anvil-makers noted in Postman, a consensus can be drawn from the comments of working blacksmiths and artisans selling their anvils that the highest quality American anvils were produced by the Hay-Budden Company of Brooklyn, NY (working dates 1890-1931). Peter Wright, the prolific English anvil manufacturer, seems to come in second in the category of most sought after anvils. While most anvil production had ceased in North America by 1970 (Bealer 1976), anvils of every description still make an appearance as a result of the exploration of New England workshops by New England's many tool pickers, collectors, and blacksmiths. Anvils are still highly prized by the many artisans seeking to perpetuate the art of the ironmonger in 21st century America.

The swage block (Figure 36) was an essential component of the workshop of many blacksmiths, as well as whitesmiths (sheet metal workers) and armorers. Squared, rectangular, circular, and other shaped holes were used to hold both the toolmakers' bar stock and associated shaping tools, which would fit into the swage block, making it a multitasking anvil for creating the multiplicity of shapes essential for toolmaking. The swage block also held the bench shears (Figure 119) and other tools for cutting, creasing, and trimming.

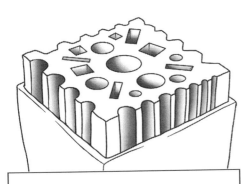

Figure 36. Swage block. Sett Balise.

Figure 37. English pattern blacksmiths' anvil. Cast steel. CLT. 22512LTC3.

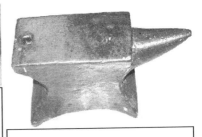

Figure 38. Horned anvil. Iron, forge-welded steel. CLT. 32313LTC1.

Figure 39. Stump anvil. Forged iron and steel, wood. DTM. 51610T1.

Leg Vises

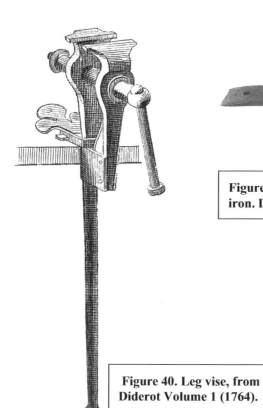

Figure 41. Post vise. Forged iron. DTM. TBF1004.

Figure 42. Leg vise. Forged iron and steel. CLT. 12813LTC2.

Figure 40. Leg vise, from Diderot Volume 1 (1764).

Mandrels

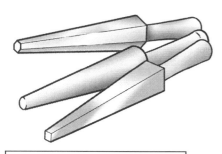

Figure 43. Hex, round and square mandrels. Steel. Illustration by Sett Balise.

Figure 44. Mandrel. Cast steel. CLT. 22813LTC10.

Hammers

Figure 45. Cross peen hammer. Cast steel, wood (hickory). CLT. 93012LTC4.

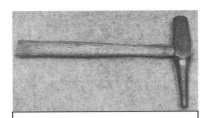

Figure 46. Blacksmiths' round eye punch hammer. Cast steel, wood (hickory). DTM. 101312T21.

Figure 47. Turning sledge. Drop-forged steel, wood (hickory). CLT. 52612LTC1.

Figure 48. Cross peen hammer. Cast steel, wood (maple). Snow & Nealley. DTM. 101312T26.

Figure 49. Dolly hammer. Forged iron, wood (hickory). DTM. 101312T5.

Figure 50. Doghead saw hammer. Steel, wooden handle. Henry Disston & Sons. DTM. 102512T7.

Figure 51. Striking sledge hammer. Cast steel, wood (hickory). DTM. 101312T15.

Figure 52. Long reach cross peen hammer. Cast steel, wood (hickory). DTM. 101312T8.

Figure 53. Blacksmiths' straight peen hammer. Steel, wooden handle. DTM. 102512T10.

Figure 54. Cold cut hammer. Cast steel, wood (hickory). CLT. 12813LTC1.

Figure 55. Silversmiths' raising hammer. Cast steel, wood (hickory). DTM. 102512T2.

Figure 56. Hammer. Cast steel, wood (hickory). DTM. 31311T7.

Tongs

Figure 57. Blacksmiths' pointed lip tongs. Forged iron. DTM. TBB1003.

Figure 58. Blacksmiths' straight fluted tongs. Forged iron. DTM. 10700T2.

Cutters

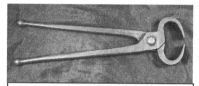

Figure 59. Blacksmiths' end cutters. Drop-forged iron and steel. Heller Brothers. DTM. 31212T1.

Figure 60. Cold chisels. Silicon manganese alloy steel. DTM. TJD1008.

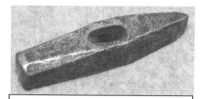

Figure 61. Cold cutter. Steel. Pinel Tool Company. DTM. 072112T4.

Figure 62. Hot set hammer. Cast steel, wood (hickory). Vulcan. DTM. 101312T6.

Figure 63. Blacksmiths' chipping hammer. Steel, wood. DTM. 102512T6.

Figure 64. Blacksmiths' end cutters. Forged steel. DTM. 2713T6.

Measuring Tools

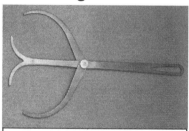

Figure 65. Blacksmiths' double caliper. Forged steel. CLT. 72812LTC1.

Figure 66. Carpenters' sliding T-bevel. Steel blade, brass trim, and rosewood handle. DTM. 31808SLP19.

Shaping Tools

Figure 67. Nail header. Natural steel. DTM. 8312T1.

Figure 68. Blacksmiths' peen hammer. Steel, wood. H.H. Harvey of Augusta, Maine. DTM. 102512T5.

Figure 69. Blacksmiths' raising hammer. Steel, wood. DTM. 102512T8.

Figure 70. Swage hammer. Cast steel, wood (hickory). CLT. 93012LTC2.

Figure 71. Anvil bottom tool for shaping. Cast steel. DTM. 2713T3.

Figure 72. Fuller hammer. Cast steel, wood (maple). Bicknell Manufacturing Co. DTM. 101312T9.

Files and Rasps

Figure 73. Blacksmiths' square file. Forged iron. DTM. 71801T8.

Figure 74. Blacksmiths' files. Drop-forged steel. McCaffrey; P.S. Stubs. DTM. 201012, 30911T2, 22411T29.

Figure 75. Farriers' rasp. Cast steel. Heller Bros. DTM. 6113T1.

Figure 76. Riffler file. Steel. Morcott & Campbell. DTM. 201012T10.

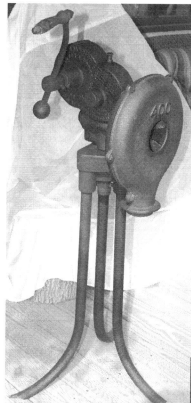

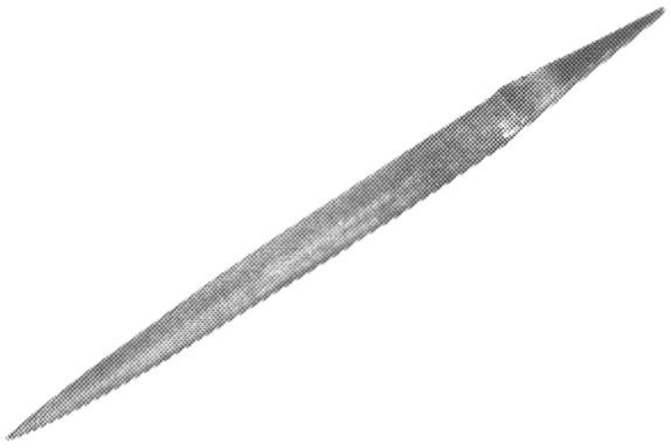

Figure 78. Crossing file. German steel. Antoine Glardon. DTM. 102012T12.

Figure 77. Blacksmiths' blower. Cast iron. Champion. CLT. 62213LTC1.

Specialized Trades of the Ironmonger

The Shipsmith

The shipsmith was a blacksmith with the special mission of making the ring bolts, chain hooks, deck irons, knee brackets, side rings, washers, and other hardware essential for the construction of a wooden ship. While the shipsmith as blacksmith had specific hardware to forge that required specific blacksmith shaping and cutting tools, his toolkit cannot be differentiated from that of the blacksmith. No documented shipsmith workshop survives from the 18th century to allow us a glimpse of how it might have differed from that of the multitasking blacksmith. Sellens (1990), who listed the tools of the shipbuilder, makes no mention of any specific tools made by the now forgotten shipsmith. In the 18th century and earlier, before trade and toolmaking specialization became widespread, the shipsmith was also often the toolmaker who forge-welded and steeled many of the edge tools used by the shipwright. In coastal communities where they worked in close proximity to a shipyard, as in Salem, Boston, Scituate, Kingston, or Old Rochester (Mattapoisett) Massachusetts, they also made other artifacts, ranging from horticultural tools to eel spears (Story 1995) and caulking irons.

Figure 79. Mortising ax. Cast or forged steel, wood. DTM. 72206T3.

Figure 80. Ship fitting. Forged iron. DTM. 10407T8.

Figure 81. Forge-welded corner chisel. Forged iron and steel. DTM. 102904T1.

Figure 82. Nail header. Steel. DTM. 102512T18.

Figure 83. Mast rings. Forged iron/steel. DTM. 12613T3.

Figure 84. Ship's block. Rope, forged iron, wood (beech, boxwood). DTM. 2713T5.

Figure 86. Shipwrights' bench clamp. Forged steel. DTM. 81713T8.

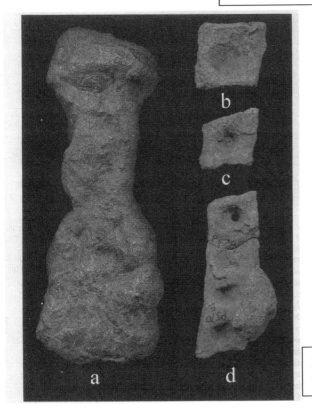

Figure 85. Caulking iron (a) found at the Fort St. George dig and dated to 1607-08 (Brain 2007, 140).

The Anchor-Smith

Perhaps just down the street from the shipsmith or at or near the local ironworks, was an anchor-smith, consolidating silicon-laced bloomery iron into the extra large bar stock necessary for anchor production. Specialized forges were used to shape these often massive anchors, which required bigger forges and large trip hammers and hoists but didn't require the finesse of the edge toolmaker. The steel anchor, including those made from galvanized and thus, rust-resistant steel, were not produced until bulk processed steel became widely available in America in the late 19th century. As with many other early tool forms, wrought iron anchors were still produced in New England at least until the Civil War. Cast iron anchors may also have been produced but would rust much more quickly than the silicon-laced wrought iron anchor, perhaps explaining why so many anchors recovered by fishing draggers from old shipwrecks off the coast of Massachusetts are obviously made from wrought iron, not cast iron (Figure 87). The obvious layers of ferrous silicate are a testament to the age, durability, and New England bog iron origin of these anchors.

Figure 87. Anchor. It is currently on display at the Davistown Museum Sculpture Garden in Hulls Cove. Cast steel. CLT. 22512LTC3. Right: Figure 88. Slitting Mill. From *Popular Science*, January 1891.

The Nail-Maker

The nail-maker was another specialized ironmonger whose slitting mill would utilize bands of rolled sheet iron from the finery or chafery to make the nail rods, which would be forged into nails.

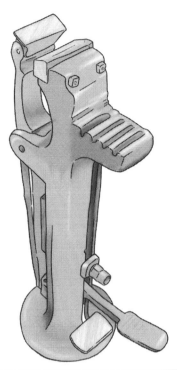

Many of the hand tools utilized by the nail-maker were smaller versions of the shipsmiths' nail headers and slitting cutters. An interesting historical footnote pertains to nail-making in America. Machinery to make cut nails instead of hand-forge-welded nails was introduced in 1795 by James Parker who, working in Newburyport, MA, patented, designed, and built a water power nail-cutting and nail-heading machine (Rosenberg 1975).

By 1798, Josiah Pierson had 96 nail cutting and heading machines operating in his slitting mill in Ramapo Village, NY (Gordon 1996, 69). In contrast, the many inventions of the English industrial revolutionaries, including nail-making machines, were not adapted in England for mass production until the second half of the 19th century. Small nail-making forges utilizing exploited teenage females working in extremely primitive and dangerous conditions are well-documented in England even as late as mid-century.

Figure 89. Farrier's Vise. Illustration by Sett Balise, based on vise by B.B. Noyse & Co. of Greenfield, MA.

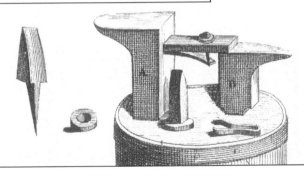 

Above: Figure 90. From Volume 2, "Nail maker," in Diderot ([1751-65] 1965a). Figure 91.

Figure 94. Above: Railroad spike. Drop-forged steel, brass. DTM. 22612T8.

Figure 92. Nail. Forged iron. DTM. 6113T6.

Figure 93. Right: Nails. Hand-forged iron. DTM. 6113T7.

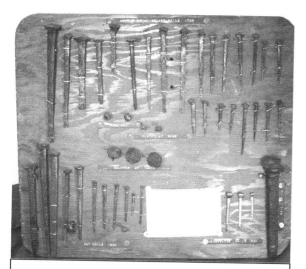

Figure 95. Nail display. Hand-forged and machine cut iron nails, wooden display. DTM. 4512T1.

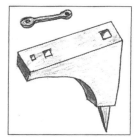

FIG. 58
Nail heading anvil, Roman 1 : 2

Taken from Book of Tools. (From Mr. John Catchings)

Nailer's anvil and header.

Figure 96. Nail anvils. From Postman (1998, 415).

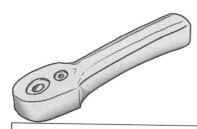

Figure 97. Nail header. Illustration by Sett Balise.

Figure 98. Bolt header. Drop-forged malleable iron and steel. DTM. 31611T5.

Figure 99. Nail and bolt header. Malleable iron. DTM. 072112T12.

The Founder

The founder, who used "draughts" and molds to shape his cast iron sleigh runners, plows, and pots, was a forerunner of the patternmakers of the Industrial Revolution. His iron pots were cast in molds that were made of clay, and later of green sand, and smoothed by the small slicks and smoothing tools, which were later used by the patternmaker to finish his molds for casting machines. Many small blast furnaces were located in the southern New England bog iron country typified by the seven furnaces documented as operating in just one Plymouth County town, Carver, after 1720 (Murdock 1937). The pigs and billets of cast iron these furnaces produced were often shipped to foundries to be melted in cupola furnaces and then further processed into products ranging from iron pots to cannonballs. By 1840, the mass production of malleable cast iron and the development of the hot blast greatly expanded the production capacity of blast furnaces and the growing repertoire of associated cupola furnaces for the specialized production of cast iron products. Some of the tools of the founder are unique creations not used by other ironmongers.

Figure 100. Founders' ladle. Forged iron. DTM. 41801T3.

Figure 101. Founders' molding tools. DTM.

The Gunsmith

The gunsmith was another specialized ironmonger, initially using malleable sheet iron to make the gun barrels. By the 19th century, high carbon steel was used for gun barrels, but because stress-related defects were present in the steel, steel gun barrels could explode upon use. As a result, advances in heat treatment technology, including austempering, allowed the development and use of ductile steel for gun production after 1875. The English industrial revolutionary Thomas Whitworth was allegedly responsible for this advance in gun-making. Gunsmiths used the ubiquitous hammer, anvil, and swedge block accompanied by the specialized simple machine, the mandrel, to shape their gun barrels.

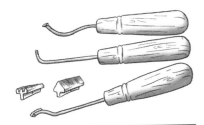

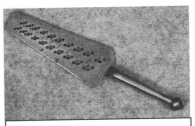

Figure 104. Horologists' and gunsmiths' screw threading plate. Cast steel. Peter S. Stubs. DTM. 42012T2.

Figure 102. Gunsmiths' gun stock tools: shovel, fluting and checkering with close-ups of checkering tool attachments. Illustration by Sett Balise.

Figure 103. Gunsmiths' screw plate. Cast steel. Peter S. Stubs. DTM. 81101T16.

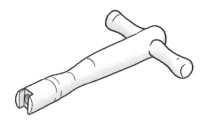

Figure 105. Gunsmiths' wrench. Illustration by Sett

Toolmakers

In the early Colonial period, the multitasking toolmaker may have been a shipsmith, gunsmith, or specialized toolmaker, such as a hammer-smith or edge toolmaker. The ubiquitous anvil and post vise would often be accompanied by the same shaping tools used by the blacksmith and gunsmith. Toolmakers forging small hand tools for any trade needed a repertoire of their own small hand tools. The small toolmakers' hand vise, a miniature version of the post vise, often made and signed by the English toolmaker Stubs, still make frequent appearances in New England tool chests, including those from rural farm workshops or the numerous workshops of metal workers of many trades. Found in every toolmakers' workshop would be the forge-welded hammer, clamps, files, and calipers that would later evolve into the factory-made drop-forged hand tools of the machinist, often made of tempered alloy steel after 1875.

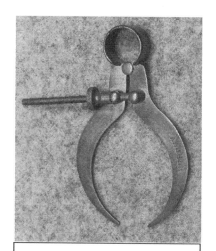

Figure 108. Adjustable wrench. Forged iron. DTM. TBK1001.

Figure 107. Hand vise. Forged iron or steel. G. Kipp. DTM. TCR1302.

Figure 106. Outside screw adjusting spring calipers. Drop-forged steel. Murphy. DTM. 62202T8.

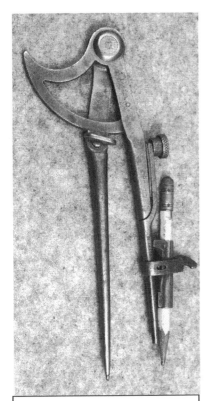

**Figure 109. Pencil compass.
Drop-forged tempered alloy
steel. W. Schollhorn. DTM.
22311T2.**

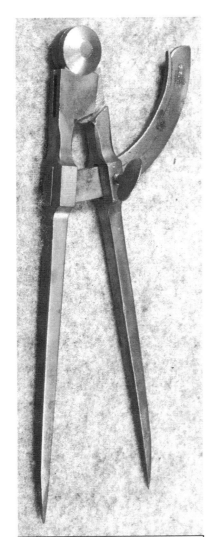

**Figure 110. Wing dividers.
Steel. G. Buck. DTM.
914108T5.**

In New England, tens of thousands of workshops were present on farms or as home
workshops from the 18th to late 20th century. These shops were characterized by the
gradual replacement of one-of-a-kind hand tools by the increasing availability of factory-
made tools. Many of the tools used in these workshops after the Civil War were drop-
forged products of the Industrial Revolution and are illustrated in *Section III*. The
contents of many 19th century workshops were often handed down by generations of
artisans to linger in 20th century workshops. These accidental durable remnants illustrate
the gradual transition from handmade hand tools to the factory-made tools that became
increasingly common after 1840. Workshops in this period continued to make handmade
tools while also repairing factory-made tools and building or repairing machinery, some
of which were one-of-a-kind creations. Among the earlier forms of hand-forged or one-

of-a-kind tools made by individual toolmakers are those illustrated below. Many a multitasking blacksmith made tools from recycled steel rasps, which when worn out, were the ideal material for tools used by farriers and other early American artisans.

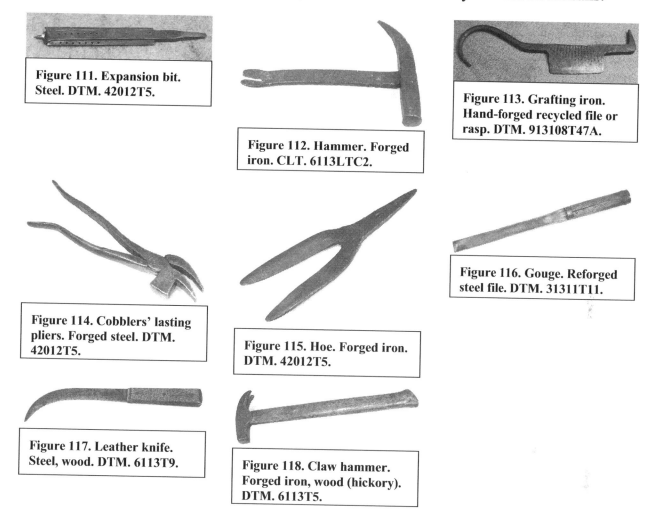

Figure 111. Expansion bit. Steel. DTM. 42012T5.

Figure 112. Hammer. Forged iron. CLT. 6113LTC2.

Figure 113. Grafting iron. Hand-forged recycled file or rasp. DTM. 913108T47A.

Figure 114. Cobblers' lasting pliers. Forged steel. DTM. 42012T5.

Figure 115. Hoe. Forged iron. DTM. 42012T5.

Figure 116. Gouge. Reforged steel file. DTM. 31311T11.

Figure 117. Leather knife. Steel, wood. DTM. 6113T9.

Figure 118. Claw hammer. Forged iron, wood (hickory). DTM. 6113T5.

The Whitesmith (Tinsmith)

One specialized metalworking trade among many early American industries was the whitesmith. The invention of galvanized sheet iron, more rust proof and durable than sheet iron, signaled the beginning of the heyday of the whitesmith. First used in Europe and England in the seventeenth century, galvanized sheet metal became widely used in America in the second half of the 18th century. Also called the tinsmith or tin knocker, the whitesmith used galvanized (tin coated) sheet iron, copper plated tin, and sheet copper to fashion many of the domestic utensils utilized at early American farms and homes, workshops, and for many commercial purposes. Tinsmiths also made the sheet metal oven for the woodstove as well as its chimney pipe, water drains, pails, teapots, and numerous other artifacts before factory-made tin containers became widely available in the late 19th century. Along with the ubiquitous tinners' snip (Figure 120), a tool that still makes an appearance in New England workshops, is the bench shear (Figure 119), which was mounted on the bench plate as a primary cutting tool of the whitesmith long before motorized cutting tools became available in the late-19th century. Also commonly associated with the whitesmith as well as with the armorer are the wide variety of stakes used to shape sheet metal, which became widely available as a result of Henry Cort's improvements in malleable (puddled) iron-producing reverbatory furnaces and rolling mills. Illustrating the transition from the handmade hand tools of early American industries to the factory-made equipment of the Industrial Revolution are the still commonly encountered cutting, shaping, and trimming sheet metal machines (Figure 130 to Figure 133) often made by the North Brothers in Berlin, CT, after 1810. Factory-made drop-forged tinsmith tools such as the hand punch, crimping tool, and levered shears depicted below illustrate the expanding repertoire of whitesmith tools in the late 19th century, many of which were used in other metalworking trades ranging from roofing to plumbing.

Figure 119. Whitesmiths' bench shears. Forged iron and steel. Peter S. Stubs. DTM. 102100T14.

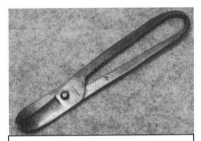

Figure 120. Curved sheet metal snips. Forged steel. DTM. 22411T20.

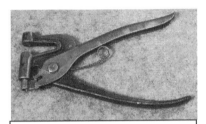

Figure 121. Tinsmiths' hand punch. Drop-forged iron and steel. DTM. 5212LTC2.

Figure 122. Bench roller with clamp and crankshaft handle. Drop-forged steel. Peck, Stow & Wilcox. CLT. 21812LTC3.

Figure 123. Tinsmiths' levered shears. Drop-forged tempered alloy steel. CLT. 3912LTC1.

Figure 124. Soldering iron. Wood, iron, copper, steel, aluminum ferrule. DTM. 31112T21.

Figure 125. Center punch. Cast steel. CLT. 8912LTC6.

Figure 126. Blowhorn stake. Cast and forged steel. CLT. 3213LTC2.

Figure 127. Soldering iron. Forged iron, copper. DTM. 3213T1.

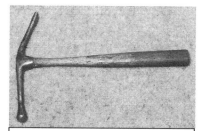

Figure 128. Coppersmiths' bumping hammer. Steel, wooden handle. DTM. 102512T9.

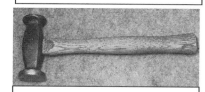

Figure 129. Brass face hammer. Cast steel, wood (hickory), brass. CLT. 10112LTC4.

Figure 130. Bench hack saw. Cast iron, wood (rosewood). Goodell Pratt. DTM. TCW1003.

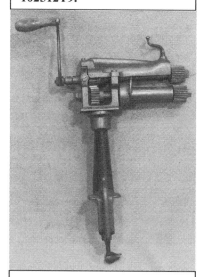

Figure 131. Bench roller with clamp and crankshaft handle. Drop-forged steel. Peck, Stow & Wilcox. CLT. 21812LTC3.

Figure 132. Top: Sheet metal crimper. Drop-forged steel, rosewood grip. CLT. 21812LTC4.

Bottom: A tinsmiths' standard mount made from drop-forged malleable iron.

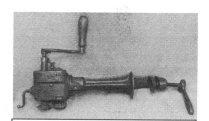

Figure 133. Bench clamped sheet metal cutter. Drop-forged steel, rosewood grip. CLT. 21812LTC5.

Figure 134. Tinsmiths' shears. Cast steel. Signed "BROWN; CAST STEEL." DTM. 81713T12.

The Farrier

Perhaps the best remembered of early American metalworkers is the farrier. While for centuries the multitasking blacksmith made ox shoes and then horseshoes as one of the many products of his workshop, by the late 18th century, the farrier was a specialized ironmonger working in both rural and urban areas. In 1829, horseshoe-making machinery was introduced by Henry Burden in his ironworks in Troy, NY (Gordon 1996, 71). The forge-welding of durable ox shoes and horseshoes from relatively high carbon malleable iron with hammer and anvil still occurred in farm workshops until the late 19th century. Among the specialized tools of the farrier are the horse rasp, toe knife (Figure 139), farriers' buttress (Figure 140), and farriers' vise.

Figure 135. Farriers' rasp. Steel, wood (rosewood), cloth tape. Nicholson. DTM. 102512T17.

Figure 136. Farriers' horseshoe creasing hammer. Cast steel, wood (hickory). H.H. Harvey. CLT. 22813LTC9.

Figure 137. Farriers' clean claw hammer. Cast steel, wood (hickory). Heller Brothers. DTM. 101312T11.

Figure 138. Buffer. Steel. DTM. 102512T17.

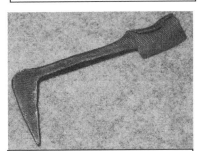

Figure 139. Farriers' toe knife. Forged malleable iron. DTM. 32912T7.

Figure 140. Farriers' buttress (hoof paring tool). Forged malleable iron, wooden handle. DTM. 61612T4.

Figure 141. Pope's iron handle farrier knife. Malleable iron handle, steel blade. T.J. Pope. DTM. 32412T6.

Figure 142. Farriers' shoeing hammer. Drop-forged steel, hickory handle. DTM. 31212T20.

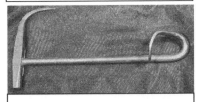

Figure 143. Farriers' snowball hammer. Forged iron and steel. H.M. Christensen of Brockton, Massachusetts. DTM. TCM1005.

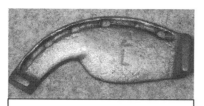

Figure 144. Ox shoe. Drop-forged steel. DTM. 32912T7.

Figure 145. Horseshoe calk wrench. Cast iron. Neverslip. CLT. 7612LTC3.

Figure 146. Clinch tongs. Forged malleable iron. Champion DeArment. DTM. 12812T5.

Figure 147. Farriers' hoof parer. Cast iron, cast steel. Champion Tool Co. DTM. 101312T28.

Figure 148. Farriers' end cut nippers. Cast iron, cast steel. Champion DeArment. DTM. 101312T24.

Figure 149. Farriers' crease nail puller. Cast steel. DTM. 101312T17.

The Edge Toolmaker

The most important of all early American ironmongers are, arguably, the edge toolmakers who forged the simple machines that felled, shaped, notched, mortised, joined, and/or carved the wood that built a nation. The toolkits of all edge toolmakers working in the pre-industrial era include the same ubiquitous simple machines of most metal-smithing trades: anvils, post vises, hammers, tongs, pinchers, cutters, clamps, calipers, and specialized shaping tools for forge-welding gouges and the sockets of chisels. Many of these tools were also used by the gunsmith, shipsmith, and specialized toolmakers.

While the specialized trade of edge toolmaking began in the English industrial centers of Newcastle, Birmingham, and Sheffield in the 17th and 18th centuries, many edge tools were forged by obscure shipsmiths working in the coastal areas of England and New England in close proximity to the shipyards that needed their iron fittings. Edge toolmaking as a specialized trade in New England began emerging with, for example, the ax-making of Richard Faxon (±1795) in or near Quincy, MA. The Quincy forges of the Faxon family were later taken over in 1824 by Jesse J. Underhill, one of the Underhill clan of edge toolmakers. The Underhill edge tool in Figure 152 is especially notable in that it clearly illustrates the inclusion of prodigious amounts of bog-iron-derived silicate slag in the body of the socket chisel .

An edge toolmaker's forge and hand tools would be hard to differentiate from that of a multitasking blacksmith. What differentiated the edge toolmaker from the shipsmith, farrier, or nail-maker was his sophisticated ability to make tool steel, which must contain at least 0.5% carbon. Without this minimum amount of carbon an edge tool cannot be properly quenched and hardened (requiring sudden cooling in brine or water), then tempered and annealed to relieve the structural stresses of the hardening process.

Early edge toolmakers, as well as famed edge toolmakers such as Thomas Witherby and the Buck Brothers, had no knowledge of the chemistry and microstructure of iron, steel, and cast iron. What made an edge toolmaker an expert at his trade was his finesse at utilizing long established rule-of-thumb procedures for forging his tools. The most important of these skills was an expertise at judging the appropriate forging temperature for a particular tool from the color of his heated steel or iron implement. At 1420 °F, cherry red, his steel would lose its body centered cubic molecular structure and its magnetism and transform itself into the face-centered cubic form, which was characteristic of austenite. Rapid cooling resulted in a harder molecular structure, as austenite became the highly stressed martensite. Tempering and annealing were the next essential steps needed to relieve the stresses characterizing the face-centered cubic form of martensite as hardened steel. The ability to execute this forge-welding coup d'état, making steel out of iron, differentiated the skilled edge toolmaker, as did his finesse at "steeling" (welding a steel cutting edge to an iron body), from the less skilled nail-maker, founder, or anchor-smith.

One of the challenges facing edge toolmakers was the successful welding of a steel cutting edge onto and under the surface of a malleable iron, wrought iron, natural steel, or lower carbon German steel tool. A close examination of many old edge tools (pre-1860) will often reveal a clear interface between the steel cutting edge and the more inexpensive iron handle. In New England, occasional examples of edge tools made of silicon-laced wrought iron appear (Figure 152). The majority of early edge tools are made entirely from natural steel by the direct process method, with evidence of forge welding throughout the tool (Figure 19). Other edge tools may involve the steeling of indirect process made decarburized cast iron (a lower carbon German steel) (Figure 31, Figure 156), which is often difficult to differentiate from malleable iron produced from the recarburization of wrought iron. Steel production in 1840 in European countries was evenly divided between indirect process German steel, incorrectly called natural steel (Day, 1991) and cementation steel from carburized wrought iron.

Cast steel, made in crucibles in small quantities and used for edge tools, razors, watch springs, and components of navigation equipment, was only a tiny percentage of total steel production at this time. Similarly, sheaf or shear steel, often of almost equal quality

54

to cast steel, was made from tediously refined cementation steel beginning in the late 17th century in England. It was the highest quality special purpose steel for almost a century before being replaced by cast steel in the late 18th century. Knowledgeable edge toolmakers, especially those trained in the European tradition of decarburizing cast iron into steel, as in the German Renaissance, continued an alternative tradition of edge tool production in the 19th century despite the ubiquitous use of cast steel for edge tools. Evidence of the continued use of German steel, which often had the advantage of an elevated manganese content, is the frequent appearance of high quality edge tools not marked "cast steel", or not obviously "steeled" in the ancient tradition of forge-welding edge tools (Figure 31).

A sequential listing of the specialized tools made by the edge toolmaker and other toolmakers in the next part of *Tools Teach* is followed by illustrations of the tools of specialized trades that used these many unique edge tools.

Figure 150. Throwing hatchet. Forged iron, steel, and wood. DTM. 61204T14.

Figure 151. Claw hatchet. Cast steel, wood. Joel Howe. DTM. TCC2011.

Figure 152. Socket chisel, forged iron and steel, signed UNDERHILL EDGE TOOL Co. DTM. 041505T24.

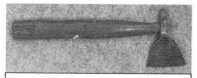

Figure 153. Belt hatchet. Reforged iron and steel, wood. DTM. 91303T2.

Figure 154. Natural steel socket framing chisel. Natural steel. DTM. 8312T3.

Figure 155. Socket framing chisel. Steel with a forge-welded edge, wood (hickory), iron ferrule. Underhill Edge Tool Co. DTM. 32707T1.

Figure 156. Framing chisel. Drop-forged iron and cast steel. A.G. Page. DTM. 31212T13.

Figure 157. Framing gouge. Cast steel, iron ferrule, hickory handle. J. Gray. DTM. 31212T15.

Figure 158. Spear head. Hand-forged malleable iron. DTM. 3312T12.

The Hammersmith

Before the iconography of the woodworking trades is illustrated, an exploration of the many forms of the hammer, a simple machine used by all metalworking and other early American trades, provides an overview of the evolution of early American industries into an Industrial Revolution characterized by a rapidly expanding repertoire of tool forms.

Just as an exploration of the iconography of edge tools leads to the many labyrinths of the cooper, wheelwright, coach-maker, cobbler, and other trades, the iconography of the hammer as that simple machine essential to the productivity of all edge toolmakers and other ironmongers needs to be depicted. As the age of early American industries evolved into the age of the American factory system, the inventory of hammers in the toolkits of artisans, mechanics, and machinists rapidly expanded as a reflection of the increasing complexity of an industrial society. The complexity of our tool kits continued to increase as we entered the Age of Petro-chemical Electrical Man with his ubiquitous automobile. For an overview of the cascading series of Industrial Revolutions that culminated in the Age of Information and Communication Technology, see Figure 743 at the end of *Part III*.

The most common hammer form in the 17th, 18th, and early 19th centuries was the multipurpose claw hammer, unchanged in design from the time of the Roman Empire until new patterns of claw hammers began appearing in the early 19th century (Figure 160).

It was David Maydole (1807-1882) who invented the adz eye hammer (1840), a drop-forged version of the ancient claw hammer designed with the intent of preventing the hammer head from falling off the hammer handle. The mass production of the adz eye hammer by Maydole, and soon by many other manufacturers, signaled the rapid rise of the American factory system of mass production. After 1840, almost all hammers were drop-forged rather than hand-forge-welded. The development of the first mass produced drop-forged hammers coincides with the development of hot blast, which greatly increased the efficiency of both the bloomery and blast furnace.

Among the most notable tools in the collection of the Davistown Museum is a 17th century hammer recovered by the Liberty Tool Company in Plymouth County, Massachusetts (Figure 5). This hand-forged wrought iron hammer was probably made by a local farmer or artisan from unrefined bog iron; it's obvious silicon filaments indicate it was made directly from a bloom of iron that had never made it to a finery to be made into iron bar stock, nor been subject to the slag expunging effects of a trip hammer. These silicon filaments echo the metallography of the wrought iron anchors made at nearby southern New England anchor forges (Figure 87). An important reference book by Baird and Comerford (1989) provides excellent illustrations of a wide variety of hammer forms, including dozens of illustrations of the hand-wrought hammers that were produced from the Roman era (or earlier) until the mass production of drop-forged hammers began after 1840.

Hammers depicted in Diderot's *Encyclopedia* ([1751-65] 1964-6) (Figure 800) illustrate

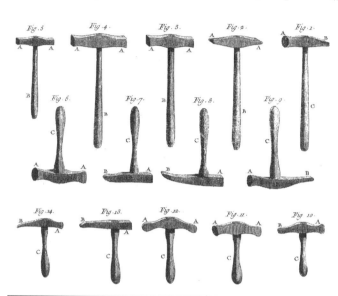

Figure 159. Section of Figure 800 from Diderot.

common European hammer forms from the late 17th to the mid-18th century, many unchanged since the early Renaissance. Those illustrated in the Diderot plates in the appendix are occasionally encountered in early American toolkits and may have been used in the same or similar trades. Since many styles of metalworking, coopers, and woodworking hammers remained unchanged for centuries, differentiating between an early 17th century and a late 18th century hammer can be difficult unless the hammer is characterized by a unique metallurgical composition. Manufacturer's marks did not appear on hammers until well after other hand tools were signed by their makers. Maydole was one of the first hammersmiths to mark his newly designed hammer. Other still commonly encountered hammer forms characterizing the early American industries discussed and illustrated in this text include those of the blacksmith, farrier, cooper, cobbler, and jeweler. Many of these are distinctly American forms, which have been adapted from European designs, including those illustrated in Diderot.

Among wooden (not iron) hammers, one form still used in the 20th and early 21st century appears in Moxon (1701), the wooden mallet. Wooden mallets are frequently encountered in many a tool chest and are commonly associated with the lighter woodworking tasks of the joiner (Figure 167 and Figure 168). Heavy duty wooden mallets were used in the shipyard by timber framers and on the farm (Figure 169). The ship caulkers' mallet (Figure 163), the production of which was perfected by Clement Drew of Kingston, MA, in the early decades of the 19th century, has antecedents going back to the early Iron Age. The illustration in Diderot ([1751-65] 1964-6) as well as those in Timmins (1976) and other English pattern books illustrate the longevity of the design of hammers and their importance for constructing British merchant men in Boston or coasting traders in many a New England port and elsewhere.

As with many other tool forms, once the American system of mass production of drop-forged tools gradually altered how tools were made, the growing multiplicity of trades and their tool forms, including hammers, expanded from the basic repertoire characteristic of early American industries. By the late 19th century, a survey of hammer forms would reveal four general categories: woodworking, metalworking, stone-working, and agricultural/domestic. Early hammer forms, such as the claw hammer, cobbler hammer, and coopers' hammer lingered and evolved into new forms.

- Woodworking: claw, veneer, cooper, marking, box, and crate
- Metalworking: blacksmith, machinist, tinsmith, shipsmith, metalsmith, jeweler, farrier, sharpening, and automotive
- Stone-working: quarrying, paving, bricklaying, and slaters
- Agricultural/Domestic: food-related (kitchen, candy, meat tenderizing), tack hammers, upholstery, leatherworking, snowball hammer, and fencing

Figure 160. Claw hammer. Forged iron and steel. Warner. DTM. 6703T18.

Figure 161. Claw hammer. Drop-forged steel, wood handle. Cheney Hammer Co., Little Falls, New York. DTM. 22612T13.

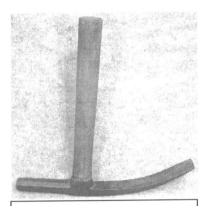

Figure 162. Cobblestone hammer. Forged steel, wood. DTM. 71401T5.

Figure 163. Caulking mallet. Wood, iron bands. DTM. 120907T5.

Figure 164. Stanley claw hammer. Cast steel, wood (hickory). Stanley. DTM. 121412T2.

Figure 165. Scotch pattern farriers' hammer. Cast steel, wood (hickory). DTM. 12613T2.

Figure 166. Rounding hammer. Drop-forged iron and steel, wooden handle. DTM. 121112T3.

Figure 167. Wooden mallet. Rosewood. CLT. 12613LTC1.

Figure 168. Wooden mallet. Hickory. DTM. 12613T4.

Figure 169. Wooden ringed maul. Wood (oak, hickory), forged iron. DTM. 12713T3.

Figure 170. Adz-eye hammer. Cast steel. DTM. 12813T3.

Figure 171. Bush hammer. Cast steel, wood (hickory). CLT. 22813LTC4.

Figure 172. Cheney nailer hammer. Cast steel, wood (hickory). Cheney Hammer Co. CLT. 10112LTC5.

Figure 173. Marking hammer. Cast steel, wood (hickory). Great Northern Paper Company. DTM. 121412T12.

Figure 174. Cobblers' beating out hammer. Steel, wooden handle, leather. DTM. 102512T11.

Figure 175. Claw hammer head. Bog iron. DTM. 81713T16.

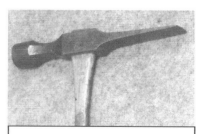

Figure 176. Carriage-makers' upholstering hammer. Drop-forged steel, wood. R. C. Clay. DTM. TJG1001A.

The Silversmith and Jeweler

The historical roots of the first silversmith predate the multitasking of the first blacksmiths who first appeared at the beginning of the Iron Age. The toolkits of both early silversmiths and coppersmiths would have been the sources of many of the designs for the tools of the blacksmith. The first complex society characterized by Royal casts can be dated to the later 4th millennium BCE. The accidental archaeological remnants of these societies, whether located in the old or new world, always seem to include evidence of the presence of the sophisticated cast of silversmiths and goldsmiths. The simple machines they utilized to produce jewelry for royalty – hammers, mandrels, dapping blocks, and holding, cutting, and measuring tools – are miniature prototypes of the simple machines of the blacksmiths, edge toolmakers, and swordsmiths who were the prime movers of Greek and then Roman society. Exploration and settlement of the New World had to wait for the invention of the gun and the evolution of cannon-producing blast furnaces. The tiny but well-to-do upper crust of Colonial society, so well illustrated by the paintings of John Copley, facilitated the florescence of a community of silversmiths who had several millennia of traditions and techniques to help execute their craft. The seemingly unimportant toolkits of the silversmith have remained unchanged in their essential makeup for centuries. Many of their hand tools are important predecessors of the toolkits of the machinist whose striking, cutting, measuring, marking, and holding tools built the machinery of the Industrial Revolution. The work of the silversmith was also impacted by an Industrial Revolution that soon resulted in the drop-forging of flatware (Figure 192 and Figure 193).

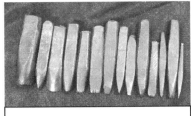

Figure 177. Chasing Stamps, forged steel. Forged steel. W. Jessop & Sons. DTM.

Figure 178. Dapping block, cast steel. DTM. 22512T9.

Figure 179. Ring clamp. Brass, wood, and leather. DTM. 41312T1.

60

Figure 180. Silversmiths' hammer. Drop-forged steel, wood (hickory). DTM. 121412T8.

Figure 181. Silversmiths' hammer. Drop-forged steel, wood (hickory). DTM. 121412T7.

Figure 182. Silversmiths' hammer. Drop-forged steel, wood (hickory). DTM. 121412T6.

Figure 183. Silversmiths' hammer. Drop-forged steel, wood (hickory). DTM. 121412T10.

Figure 184. Silversmiths' hammer. Drop-forged steel, wood (hickory). DTM. 121412T9.

Figure 185. Draw plate. Cast steel. DTM. TCP1003.

Figure 186. Pin vise. Drop-forged steel, brass. DTM. 22612T7.

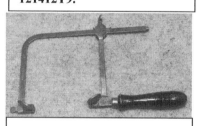

Figure 187. Jewelers' coping saw. German steel, steel blade, hardwood handle. DTM. 22211T24.

Figure 188. Screw threader. Cast steel. Bleckmann. DTM. 32912T3.

Figure 189. Silversmiths' hammer. Cast steel. DTM. 51311T1.

Figure 190. Silversmiths' hammer. Cast steel. DTM. 2713T1.

Figure 191. Silversmiths' forming tools. Cast steel. DTM. 32413T1.

Figure 192. Silver flatware molds. Towles Silversmiths. Cast steel. CLT. 6113LTC1.

Figure 193. Silver flatware molds. Towles Silversmiths. Cast steel. CLT. 6113LTC1.

Woodworking Tools

In the decades and centuries before the Industrial Revolution, the iconography of hand tools begins with the simple machines of the blacksmith, shipsmith, farrier, gunsmith, whitesmith, and edge toolmaker, and his or her hammers, anvils, bench plates, and swage blocks. In England, by the later 17th and throughout the 18th and early 19th centuries, families of toolmakers gathered at industrial centers such as Birmingham and elsewhere to forge-weld the hand tools that would build an empire. Families, clans, and communities of toolmakers began working in New England as early as the great migration to Boston (1629-1643). By the late 18th century, especially after the American Revolution, New England's network of rivers began powering a vigorous toolmaking community, supplying the water power they needed for furnaces, forges, and trip hammers. The most common use of the tools they forge-welded was for woodworking with hunting and warfare arguably a close runner-up. All woodworking trades included one or more of the following functions: felling, hewing, shaping, mortising, notching, boring, planing, and measuring.

A logical sequence of arranging the iconography of woodworking tools closely follows the logical sequence of their use. The felling, hewing, and transport of timber from forest to coasting vessel or oxen yoked sledge preceded the shaping of this timber with broad ax, side ax, slick, pit and crosscut saws. After the felling of timber, specialized tools such as the spud were used to remove the bark. Timber scribes were then used to mark an owner's initials. Timber framing and ship building also required mortising and notching, not to mention boring for trunnels and/or ring bolts and further special purpose shaping. Specialized woodworking trades soon supplemented those of the timber harvester, timber framer, and shipwright. The joiner, for example, a ships' cabinetmaker, was soon joined by the cooper, often a farmer, as shingle, shake, and barrel-maker, and other specialized tradesmen, such as spar-maker, coach-maker, wheelwright, and coffin-maker. The

62

furniture maker is an ancient trade that also frequently employed the tools of the cabinetmaker. Distinctive tool forms also often characterize the toolkits of all specialized woodworking trades.

Group 1: Felling and Hewing Tools

The all purpose English felling and shaping ax soon evolved into specialized ax designs suitable for America's forests where huge softwood trees (e.g. white pine and birch) were interspersed with hardwoods essential for strong ship frames (white oak, beech, maple). The large variety of ax forms that had evolved by the late 19th century is well illustrated in Kauffman (1966), Heavrin (1998), and Klenman (1990). Edge tools dominate all woodworking toolkits but are always accompanied by boring tools, planes, measuring tools such as mortising, panel, depth, miter, and bevel gauges and framing squares. Many an underlying saw form can be traced from the roman era through Moxon ([1703] 1989), Diderot ([1751-65] 1964-6), and the English pattern books (Smith [1816] 1975) to the multiplicity of forms that appear in the tool catalogs of American saw-makers in the late 19th century (See *Section III*).

Figure 195. Hewing ax. German or blister steel. DTM. 111006T2.

Figure 196. Hewing ax. Forged iron and steel. DTM. TAX3500.

Figure 194. Double bit ax. Cast steel, wood (hickory). Plumb. DTM. 12813T2.

Figure 198. Peavey. Drop-forged iron and steel, wooden handle. Peavey Mfg. Co. DTM. 4106T9.

Figure 197. Lumber pick. Low carbon steel. DTM. 93011T6.

Figure 199. Pickaroon. Forged iron, natural steel, wood. DTM. 102904T5.

Figure 200. Pulp hook. Drop-forged iron and steel, wood (rosewood). Snow & Nealley. DTM. 31908T39.

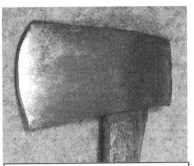

Figure 201. New England pattern ax. Forge-welded iron and cast steel, wood. DTM. 091909T2.

Figure 202. Spud. Malleable iron and steel, wood. DTM. 31808PC6.

Figure 203. English felling ax. Forged steel, wood (hickory). DTM. 81812T1.

Figure 204. Single bit hewing ax. Forged iron and steel, wood. DTM. 42012T1.

Figure 205. Felling ax. Iron, steel, wood. W. Griffin. DTM. 072112T3.

Figure 206. Cant dog. Forged malleable iron, wood. DTM. 7309T7.

Figure 207. Tweaker lumber tool. Drop-forged steel. DTM. 101312T10.

Figure 208. Pickaroon. Forged steel, wood (rosewood). DTM. 121412T17.

Figure 209. Loggers' board rule. Boxwood and brass. Stanley No. 81. DTM.

Figure 210. Bark spud. Forged iron. DTM. TCO1002.

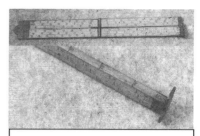

Figure 211. Folding log measure rule. Boxwood, brass. Chapin-Stephens. CLT. 71512LTC1.

Figure 212. Maine pattern double bit ax. Forge-welded iron and steel, hickory wood handle. DTM. 6112T1.

Figure 213. Log rule. Wood. DTM. 71903T7.

Figure 214. Log caliper. Boxwood and brass. Valentin Fabian. DTM. 51102T1.

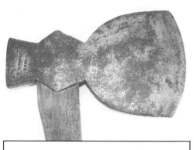

Figure 215. Yankee pattern ax. Cast steel, wood. DTM. 102612T1.

Figure 216. Yankee pattern mast ax. Cast steel, wood. DTM. 102612T2.

Figure 217. Walking wheel caliper. Wood, brass. LTC. 12812LTC1.

Figure 218. Race knife or timber scribe. Forged steel, brass. CLT. 42112LTC1.

Figure 219. Race knife or timber scribe. Cast or forged steel and wood. DTM. 101400T17.

Figure 220. Ax. Snow & Neally. Cast steel, wood (hickory). CRF.

Figure 221. Peavey. Forged steel, wood (hickory). CLT. 22813LTC2.

Figure 222. Two-man wood saw. Spring steel, iron, wood (beech). DTM. 12713T1.

Figure 223. Virginia pattern crosscut saw. Saw steel, wood (rosewood). DTM. 2213T1.

Group 2: Shaping, Mortising, and Notching Tools

Figure 224. Timber framing socket chisel. Forged iron and cast steel, wood. Underhill Edge Tool Co. DTM. 93011T7.

Figure 225. Mortising chisel. Drop-forged iron and cast steel, wood. H.T. Blodget. DTM. 22512T12.

Figure 226. Mortising chisel. Cast steel, wood (beech). J. Russell. DTM. 3213T3.

Figure 227. Corner chisel. Forged steel. G. Sheldon. DTM. 6703T1.

Figure 228. Shingling hatchet. Forged iron and steel, wood. Underhill Edge Tool Co. DTM. TCC3003.

Figure 229. Mortising chisel. Malleable iron and cast steel. James McWarth. DTM. 31212T18.

Figure 230. Carpenters' socket gouge. Drop-forged steel. A.W. Crossman. DTM. 32412T5.

Figure 231. Mortising ax, forge-welded iron and steel. Used for timber framing. DTM. 72206T2.

Figure 232. Carpenters' outside bevel gouge. Forged iron, steel, and wood. Horton, New York. DTM. 090109T3.

Figure 233. Gutter adz. Forged iron and steel, wood. DTM. 090105T1.

Figure 234. Drawknife. Reforged steel file, hardwood handles, brass ferrules, wood. DTM. 32412T2.

Figure 235. Mortising ax. Malleable iron. DTM. 102612T13.

Figure 236. Inside bevel gouge. Cast steel, wood (hickory). J. Gray. DTM. 31212T16.

Figure 237. Carpenters' drawknife. Drop-forged steel, wood. DTM. 22512T6.

Figure 238. Yankee pattern ax, sometimes called a side ax. Forged iron and natural (?) steel with wooden handle. H. Bragg of Cornville. DTM. 062603T1.

Figure 239. Ax. Forged steel, wood (oak). H. N. Dean. DTM. 32313T3.

Figure 240. Ax. Forge-welded steel, hickory wood handle. Emerson & Stevens of Oakland, Maine. DTM. 6112T2.

Figure 241. Pit saw. Wood, forged saw steel. DTM. 1302T1.

Figure 242. Carpenters' froe. Forged iron and steel. DTM. TCS1001.

Figure 243. Gutter hand adz.
Forged steel, wood (hickory).
DTM. 101312T16.

Figure 244. Wedge. Cast steel.
DTM. 121412T14.

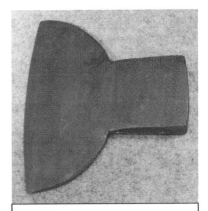

Figure 245. Broad ax. Forged
iron, steel. DTM. 41203T13.

Group 3: Boring tools

Figure 246. Tapered auger bit.
Cast steel. F. Walker of
Sheffield. DTM. 31602T5.

Figure 247. Auger bit. Forged
iron, wood. J.T. Pugh of
Philadelphia. DTM.
TCE1003A1.

Figure 248. Pipe or pump
auger bit. Forged iron. DTM.
TBA1004.

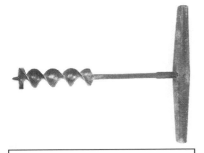

Figure 249. Carpenters' auger.
Forged iron and steel, wood
handle. C. Drew. DTM.
111002T1.

Figure 250. Center bit set. Cast
Steel. Various makers. CLT.
63012LTC1.

Figure 251. Bow drill. Lathe
turned steel, leather, wood.
DTM. 8912T1.

Figure 253. Snail countersink bit. Wood (rosewood), cast bronze, steel cutters. DTM. 81212LTC6.

Figure 254. Diamond center gimlet bit. Cast steel. DTM. 8912T8.

Figure 252. Screwtip auger bit. Forged iron. DTM. 21013T1.

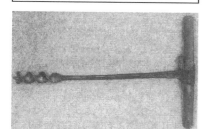

Figure 256. Spade bit. Wrought or malleable iron. DTM. 111412T2.

Figure 255. Ring auger. Hand-forged steel, wood. T. Snell. DTM. 72712T1.

Figure 257. Expansive bit. Drop-forged steel. W. A. Ives Mfg. Co. DTM. 111412T1.

Figure 258. T-handle nut auger. DTM. 22813T1.

Figure 259. Bit brace. Drop-forged grey cast iron and steel. The Jacob Mfg. Co. DTM. 22601T1.

Figure 260. Continuous motion two-speed ratcheting breast drill. Cast iron, steel, wood (rosewood). Millers Falls Co. CLT. 32313LTC6.

Group 4: Planes

All planes were handmade before the appearance of Hazard Knowles first cast iron planes (1825). One-of-a-kind cast brass and bronze planes were predecessors of and sometimes prototypes for later factory-made cast iron and steel planes. Wooden planes made before 1800 were usually characterized by a heavily chamfered edge. Beech replaced birch as the wood used to make planes after 1800. Planes made by shipwrights often used tropical woods such as rosewood, lignum vitae, and cocobolo derived from the

West Indies trade and the forests located inland along the Bay of Campeachy. Illustrations of trade-specific planes follow this introduction to early planes. A more detailed listing of factory-made planes, including those made by the Stanley Tool Co., is in *Section III*.

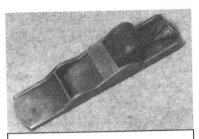

Figure 261. Brass low angle block plane. Cast bronze with cast steel blade. DTM. 22512T4.

Figure 262. Brass scraper. Cast bronze body, steel blade, malleable iron cap. DTM. 22612T3.

Figure 263. Bronze floor plane. Cast bronze, forged steel blade. CLT. 6912LTC2.

Figure 264. Spar plane, wood (beech), cast steel blade, signed J. R. TOLMAN HANOVER MASS on body and WILLIAM ASH & CO on blade. DTM. 42904T7.

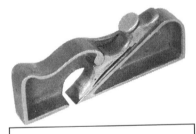

Figure 265. Brass rabbet plane. Cast bronze, steel cutter. CLT. 81212LTC14.

Figure 266. Cast brass scraper. Cast bronze, steel edge. CLT. 81212LTC12.

Figure 267. Beading plane. Wood (beech), steel blade. P.M. Peckham, Fall River. DTM. 1302T5.

Figure 268. Fore plane. Wood body, Swedish steel cutting edge. Bernard and Oskar Liberg of Rosenfors, Sweden. DTM. 102512T23.

Figure 269. Complex molding plane. Wood (yellow birch), steel blade. John Flyn. DTM. 71504T3.

Figure 271. Coffin plane. Rosewood and maple with cast steel blade. DTM. 81801T10.

Figure 272. Panel raising plane. Wood (beech), cast steel blade. Abiel F. Walker. DTM. 92001T2.

Figure 270. Double sash plane. Wood (beech), steel blade. DTM. 31808SLP5.

Figure 273. Rabbet plane. Lignum vitae with an oak wedge and a steel blade. DTM. 32802T8.

Figure 274. Bench plane or fore plane. Wood (birch). Levi Tinkham. DTM. TCD1003.

Figure 275. Panel raising plane with adjustable fence. Wood (beech), steel blade. James Cam. DTM. TCD1003.

Figure 276. Tongue and groove plane. Oak and brass. Abiel F. Walker. DTM. 101801T2.

Figure 277. Wood bottom transitional jointer plane. Wood (beech) body and handles, cast iron fittings. Birmingham Plane Co. CLT. 72312LTC4.

Figure 278. Panel raising plane. Wood (beech). Thomas Waterman. DTM. TBW1003.

Figure 279. Plane. Wood (beech), steel. Thomas Waterman. DTM. TBW1004.

Figure 280. Rabbet plane. Wood (beech), steel. Joseph Metcalf. DTM. TBW1009.

Figure 281. Horned scrub plane. Wood (beech, hornbeam), steel, brass. E.C. Emmerich. CLT. 2713LTC2.

Figure 282. Beading plane. Wood (beech), steel. Joseph Bushell Bridge. CLT. 3713T3.

Figure 283. Moving filletster plane. Wood (beech), steel. James Clark. CLT. 3713T4.

Figure 284. Plow plane. Wood (beech, boxwood), steel. A.B. Semple & Bros. CLT. 32313LTC2.

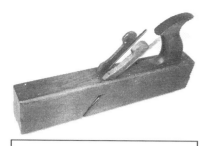

Figure 285. Toted double cutter door plane. Wood (beech), cast steel. Levi Tinkham. DTM. 32313T1.

Figure 286. Complex molding plane. Wood (beech), cast steel. CLT. 32313LTC9.

Figure 287. Complex molding plane. Wood (beech), cast steel. W. King. CLT. 32313T2.

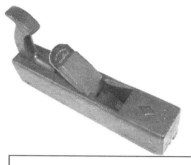

Figure 288. Toted match plane. Wood (maple), cast steel. DTM. 32313T5.

Group 5: Measuring Tools

Figure 289. Bevel. Wood (rosewood). CLT. 72712LTC2.

Figure 290. Caliper chisel. Cast steel, wood (rosewood). Buck Brothers. DTM. 121412T11.

Figure 291. Carpenters' steel point dividers. Tempered alloy steel. Joshua Stevens Arms & Tool Co. DTM. 32912T5.

73

Figure 292. Plumb bob. Brass. C. Drew & Co. CLT. 51312LTC1.

Figure 293. Stanley no. 89 sliding clapboard gauge. Stanley. DTM. 101312T30.

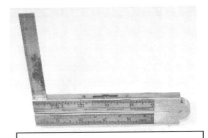

Figure 294. Combination rule. Stephens & Co. CLT. 3213LTC3.

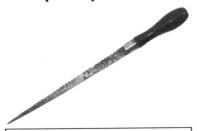

Figure 295. Level. Wood (rosewood), brass, glass. James M. Davidson. DTM. 32313T6.

Group 6: Specialized Woodworking Trades

Figure 296. Planemakers' float. German steel, brass ferrule, hardwood handle. DTM. 111412T16.

Figure 297. Mallet. Horn and wood. DTM. 22311T10.

Figure 298. Veneer saw. Cast iron, cast steel, lacquered wood. DTM. 101312T27.

Figure 299. Stanley No. 85 razor edge spokeshave. Boxwood, drop-forged steel. CLT. 12413LTC1.

Figure 300. Spokeshave. Wood (rosewood), cast steel. CLT. 12413LTC2.

Figure 301. Planemakers' float. Cast steel, wood. DTM. 31212T14.

Tools of the Woodworker by Trade

The prime movers of the ancient civilizations of the Bronze Age were wind and wheels. In the Iron Age (>1200 CE), iron smelting furnaces created the iron blooms, which after forging into bar stock, were the raw materials used by the shipsmith to forge his edge tools and ship fittings. The coming of the steam engine signals the advent of a new Industrial Revolution that took a century and a half to obliviate the wind-powered Wooden Age. The last half century of the Wooden Age saw 4, 5, and 6-masted wooden schooners efficiently transporting the multiplying products of cascading Industrial Revolutions. Iron and steel were used not only for edge tools and ship fittings, but for machine engines and machinery production. Steam –powered railroads soon transported the majority of the products of our new factory system of manufacturing. Woodworking tools lingered in the shadows of our cascading Industrial Revolutions, yet still have relevance in the contemporary age of klepto-plutocracy. The looming possibility of collapsing electronic grids and the infrastructure of global consumer society due to geomagnetic solar storms adds to the uncertainties of the Age of Information Technology. Woodworking tools may yet emerge again as essential components of convivial, alternative sustainable economies. Some of the most interesting woodworking tools in the collection of the Davistown Museum, as well as commonly encountered tools in our Tools Teach program, are depicted in the following sections of this text.

The description of categories of woodworking tools by function needs to be followed by an iconography of woodworking tools by trades. As with many of the tools used by multi-tasking ironmongers, common tools were used in many different trades. The drawknives, try squares, and saws of the carpenter could have been used in any specialized trade from cabinetmaker to shipwright or coffin-maker. Yet some woodworking tools have unique designs, some subtle and others trade-specific, that make them commonly encountered artifacts in the toolkits of a particular woodworking trade. The lipped adz, for example, would usually be found in the tool chest of a shipwright and not a cabinetmaker.

Carpenter

Figure 302. Timber framing chisel. Forged iron, cast steel, wood. Buck Brothers. DTM. 31908T20.

Figure 303. Carpenters' keyhole saw. Forged steel, beech. DTM. 32412T1.

Figure 304. Carpenters' try square. Steel scale, brass, rosewood handle. Winchester. DTM. 22311T14.

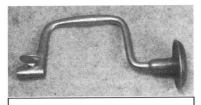

Figure 305. Carpenters' folding drawknife. Forged iron and steel, wooden handle. A.J. Wilkinson. DTM. 52403T4.

Figure 306. Carpenters' back or tenon saw. Cast steel, wooden handle. I. Colbeck. DTM. 92911T17.

Figure 307. Hand saw. Cast steel, solid brass, wooden handle. B. Biggin & Sons. DTM. 072112T2.

Figure 308. Spofford bit brace. Drop-forged iron. John S. Fray & Co. CLT. 42912LTC18.

Figure 309. Hand saw. Cast steel, solid brass, wooden handle. B. Biggin & Sons. DTM. 072112T2.

Figure 310. Scratch gauge. Wood (ebony), brass, steel. Robert Fairclough & Co. CLT. 3713T5.

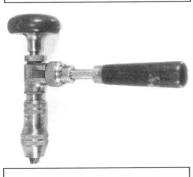

Figure 311. Stanley No. 984 ratcheting corner bit brace. Nickel-plated steel frame with wood (cocobolo) handles. CLT. 81212LTC8.

Figure 312. Drawshave. Cast steel, wood (hornbeam). Snow & Nealley. DTM. 121412T1.

Figure 313. Stanley No. 95 Butt gauge. Grey cast iron, nickel plate finish. CLT. 122712LTC1.

Figure 314. Corner chisel. Cast steel. T.H. Witherby. CLT. 101212LTC3.

Figure 315. Adz hammer. Cast steel, wood (hickory). Cheney Hammer Co. CLT. 10112LTC6.

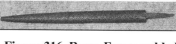

Figure 316. Rasp. Forge-welded German steel. DTM. 42012T3.

Figure 317. Drawknife. Steel, wooden handles. J.P. Davis. DTM. 102612T8.

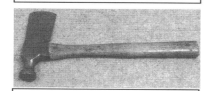

Figure 318. Drawknife. Steel, wooden handles. C. J. Kimball. DTM. 102612T9.

Figure 319. Rasp. German steel, brass ferrule, wooden handle. DTM. 102512T19.

Figure 320. Carpenters' clapboard slick. Forged iron and steel, wood handle. Billings, Augusta. DTM. 52403T3.

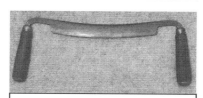

Figure 321. Shingle hatchet. Drop-forged steel, wooden handle. C.A. Williams & Co. DTM. 43006T9.

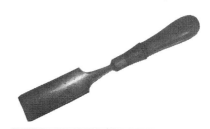

Figure 322. Outside bevel gouge. Cast steel. Holland & Turner. DTM. TCC2002.

Figure 323. Slick. Forged iron and steel, brass ferrule, wooden handle. Oliver & Daniel Babcock. DTM. 32113T1.

Figure 324. Timber framing chisel. Forged iron and steel, steel ferrule, wooden handle. Dean & Sawyer. DTM. 32113T2.

Figure 325. Mortising chisel. Cast steel, forged steel, wood (hickory), brass. William Ash & Co. DTM. 32313LTC5.

Cabinetmaker and Joiner

Figure 326. Carpenters' auger bits. Drop-forged steel, wood. Russell Jennings & Stanley. CLT. 3312LTC1.

Figure 327. Multi-tool handle. Rosewood handle with drop-forged malleable iron or steel bits. John S. Fray. DTM. 22311T3.

Figure 328. Gentlemans' bit brace. Brass, wood. A.&W. Jinkimson of Sheffield. DTM. 32708T56.

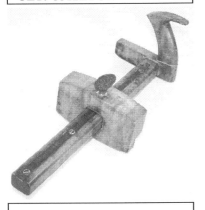

Figure 329. Slitting gauge. Wood, forged iron. DTM. 42801T11.

Figure 330. Sliding mortise gauge. Wood, brass. DTM. 31808SLP27.

Figure 331. Miter square. Forged steel, brass screws, rosewood. Stanley. CLT. 6212LTC2.

Figure 332. Level. Wood (rosewood), brass. Stratton Brothers. DTM. 13102T1.

Figure 333. Tenon saw. Tempered alloy steel, brass, and wooden handle R. Groves & Son. DTM. 92911T16.

Figure 334. Millers Falls cigar spokeshave. Cast steel, wooden (rosewood) handles. Millers Falls Co. CLT. 7712LTC6.

Figure 335. Carpenters' folding rule. Ivory, German silver. Stephens & Co. CLT. 81212LTC2.

Figure 336. Scratch gauge. Tool steel. DTM. 8312T6.

Figure 337. Cabinetmakers' keyhole saw. Brass, wooden (beech) handle. CLT. 7712LTC8.

Figure 338. Cabinetmakers' dovetail saw. Cast steel, brass fuller, wooden (beech) handle. William Marples. CLT. 7712LTC7.

Figure 339. Butt and rabbet gauge. Brass, wood (rosewood). CLT. 122712LTC2.

Figure 340. Cabinetmakers' try square. Drop-forged tempered alloy steel, brass, rosewood. DTM. 3312T14.

Figure 341. Screw box. Wood and steel. DTM. 112303T1.

Figure 342. Tapered tenon saw. Cast steel, brass, and wooden handle. R. Groves & Son. DTM. 92911T15.

Figure 343. Drawknife. Forged iron and steel with laminated edge, brass, wood (rosewood) handles. J. T. Coffin & Son. DTM. 12812T4.

Figure 344. Cabinetmakers' drawknife. Drop-forged steel, wood, painted black. Charles E. Jennings. DTM. 31808SLP17.

Figure 345. Try square. Tempered alloy steel, brass, wood. Disston & Morrs. DTM. 121112T4.

Figure 346. Keyhole saw. Cast steel, brass, wood. William McNiece of Philadelphia. DTM. 121112T5.

Furniture-Maker

Figure 347. Dowel pointer. Malleable cast iron, steel. DTM. 22311T11.

Figure 348. Adjustable dowel and spoke pointer or hollow auger. Cast iron, steel. CLT. 52712LTC1.

Figure 349. Adjustable hollow auger bit. Malleable cast iron, steel. E.C. Stearns. DTM. 101312T23.

Figure 350. Hand router. Cast bronze, steel cutter. CLT. 81212LTC16.

Figure 351. Hand beader. DTM. 82512T1.

Figure 352. Windsor beader. DTM. 81212LTC6.

Shipwright

Figure 353. Timber framing chisel. Malleable iron, steel, with forge-welded laminated steel. Gary I. Mix. CLT. 22512LTC13.

Figure 354. Slick. Forged cast steel, rosewood. Cobb & Thayer. Courtesy of Frank Kosmerl. 6712LTC2.

Figure 355. Peen adz. DTM. 102612T6.

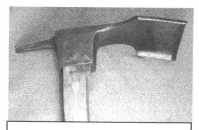

Figure 356. Slick. Forge-welded steel, wood (oak). DTM. 121412T5.

Figure 357. Slick. Forged iron and steel. T.C. Jackson of Bath, Maine. DTM. 121112T1.

Figure 358. Lipped adz. Formerly in the collection of George DuPrey. Drop-forged cast steel, hardwood handle (oak?). Collins & Co. DTM. 21812T20.

Figure 359. Lipped adz. Steel, wooden handle. J.P. Billings of Clinton, Maine. DTM. 121906T2.

Figure 360. Peen adz. Forged iron and steel, wooden handle. J. Stuart. DTM. 12712T2.

Figure 361. Peen adz. Forged iron, cast steel, wood. Boston Arnold. DTM. 72801T4.

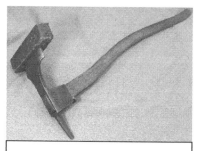

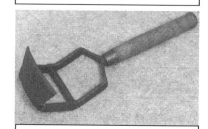

Figure 363. Gutter adz. Forged steel and iron, wooden handle. Jim Sheffield. DTM. 121112T2.

Figure 362. Lipped adz. Formerly in the collection of George DuPrey. Cast steel, hardwood handle (oak?). Plumb. CLT. 21812LTC1.

Figure 364. Ship scraper. Manufactured by Clement Drew. Kingston, MA. DTM. 22411T1.

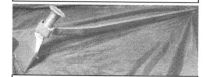

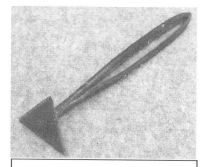

Figure 365. Hawsing iron. Forged malleable iron and steel. DTM. 012705T1.

Figure 366. Ship scraper or deck scraper. Forged malleable iron. CLT. 72612LTC1.

Figure 367. Ship carpenters' scraper. Forged iron, steel, wood. DTM. 71401T9.

Figure 368. Razee plane. Wood (beech), cast steel blade. C.A. Spear. DTM. 080704T1.

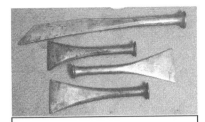

Figure 369. Caulking iron set. Cast steel. DTM.

Figure 370. Shipwrights' maul. Forged malleable iron, wood (ash, hickory). 102012T2. Courtesy of Sett Balise.

Figure 371. Caulking irons. Cast steel. T Laughlin & Co. CLT. 22813LTC3.

Cooper

If we could sketch the interrelationship of the early American industries essential for the survival and growth of a mercantile economy we would, of course, begin with the blacksmiths and woodworkers whose tools are the prime movers of early American industries. The tools of all specialized trades would be components of this diagram but something would be missing: the agricultural implements of the farmer and the unique tools of the fisherman, most of which were made domestically and not imported from England. It was cod fish and agricultural products that were initially essential for the survival of our ironmongers and shipwrights, none of whom could survive without food, water, and beer. Herein lays the hidden role of the cooper. The products and artifacts made by the cooper, who were both male and female, were the key to the transport and storage of agricultural products as well as the salt, rope, water, beer, wine, and food products that were the essential components of any successful fishing and/or trading expedition. The tools used by the also multi-tasking cooper have unique forms, which give expression to their unique uses.

The cooper and the practical implements he/she created, is an important component of the iron/wood/food triad, which is fundamental to both the survival of the first colonists and the evolution of a successful mercantile trading economy. Without the natural resources characteristic of North America (furs, forests, and fisheries) and the relatively simple toolkits of the early industries of Colonial America, both the conquest of First Nation communities and achieving independence from England would have been impossible. Other early American artisans – sawyers, coach-makers, wagon-makers, wheelwrights,

82

and patternmakers supplement the production of the timber framers, shipwrights, cabinetmakers, and carpenters of a Wooden Age that flourished until the mid-19th century, cascading Industrial Revolutions notwithstanding. The cooper was one of the obscure prime movers of the colonial and early American maritime economy. His or her products were essential components of the commercial activities of a wind- and wood-powered economy.

Figure 372. Shingle knife. Puddled or German steel, wood handle. Higgins & Libby of Portland, Maine. DTM. 121311T2.

Figure 373. Coopers' broad ax. Forged iron and weld steel with wood handle. H.A.W. King. DTM. 7602T2.

Figure 374. Croze. Wood, cast steel, and forged iron. DTM. 100400T7.

Figure 375. Coopers' down shave. Wood with a forged or cast steel blade. Kenyon, Sheffield. DTM. TCJ1001.

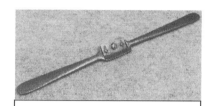

Figure 376. Coopers' drawshave. Forged malleable iron and steel. DTM. 32412T3.

Figure 377. Coopers' hoop driver. Drop-forged steel, hickory. DTM. 22512T2

Figure 378. Howel. Wood and steel. DTM. 42801T7.

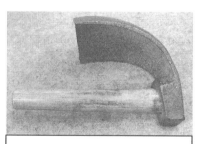

Figure 379. Bowl adz. Forged iron and steel, wood. DTM. 90608T1.

Figure 380. Coopers' shave. Forged steel with welded edge. DTM. 72812T6.

Figure 381. Coopers' hollowing drawknife. Forged steel, wood. Snow & Nealley. DTM. 31212T3.

Figure 382. Coopers' adz. Malleable iron and steel with wood handle. DTM. 7602T3.

Figure 383. Stanley No. 57 coopers' shave. Cast grey iron with steel blade. DTM. 040103T2.

Figure 384. Coopers' sun plane. Wood (bird's eye maple), cast steel blade. Greaves & Co. DTM. 62212T3.

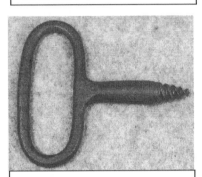

Figure 385. Coopers' vise. Forged iron. DTM. 62212T2.

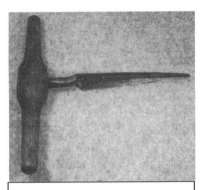

Figure 386. Coopers' tap borer. Forged iron and steel, wood. DTM. TCE1001.

Figure 387. Hoop set hammer. Cast steel, wood (maple). DTM. 101312T14.

Figure 388. Coopers' bung borer. Forged iron and steel, wood. DTM. 100400T17.

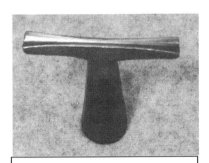

Figure 389. Hoop driver. DTM. Cast steel. 72812T2.

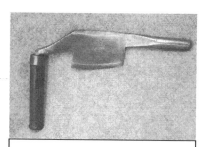

Figure 390. Coopers' chamfer knife. Drop-forged iron and steel, wood. Leonard & Ichabod White. DTM. 41203T6.

Figure 391. Coopers' drawshave. Cincinnati Tool Co. Cast iron, steel, wood (oak). CLT. 82512LTC1.

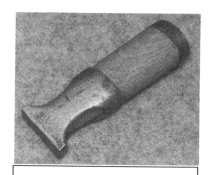

Figure 392. Hoop set. Cast steel, cast iron, wood (maple). Sloan. DTM. 101312T20.

Figure 393. Coopers' flagging iron. Malleable iron and steel. DTM. 102612T12.

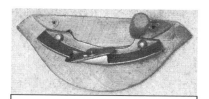

Figure 394. Coopers' croze. Oak body, steel edge. DTM. 102512T22.

Figure 395. Coopers' bung. Wood, steel ferrules. DTM. 102512T21.

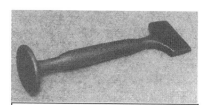

Figure 396. Hoop driver. Cast steel. DTM. 102512T20.

Figure 397. Coopers' bung hammer. Steel, wooden handle. Lang & Jacob. DTM. 102512T3.

Sawyer

Figure 398. Saw set. Drop-forged steel. Nike of Eskiltuna, Sweden. DTM. 22311T13.

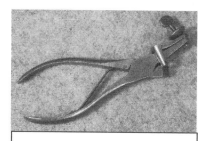

Figure 399. Saw set. Drop-forged steel with hardened steel jaws. Bemis & Call. DTM. 30202T2.

Figure 400. Saw set. Drop-forged iron. CLT. 42912LTC7.

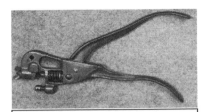

Figure 401. Crosscut saw set. Cast iron. CLT. 6312LTC10.

Figure 402. Saw gauge. Drop-forged steel. DTM. 72801T18.

Figure 403. Saw gauge. Cast steel, cocobolo. CLT. 32512LTC1.

Figure 404. Saw swage. Drop-forged steel. Simonds. DTM. 102503T4.

Figure 405. Hand saw. Spring steel. Wheeler, Madden & Clemsen. DTM. 12813T1.

Figure 406. Back saw. Spring steel. Wheeler, Madden & Clemsen. DTM. 12813T1.

Figure 407. Carpenters' saw. Saw steel, wood. DTM. 102612T14.

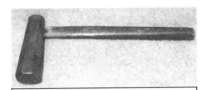

Figure 408. Saw hammer. Steel, wooden handle. Fisher. DTM. 102612T7.

Figure 409. Saw set. Drop-forged steel. Atkins. DTM. 102512T13.

Figure 410. Hand saw. Saw steel, wood (rosewood), brass. T. Tillotson. DTM. TCW1003.

Figure 411. Hand saw. Spring steel, wood (beech). R. Groves & Sons, Sheffield. DTM. 2213T2.

Coachmaker/Wagon-Maker/Wheelwright

Figure 412. Wheelwright's wooden compass plane. Levi Tinkham. Beech wood, steel. DTM. 51213T1.

Figure 413. Wagon jack. Cast steel, wood (hickory). DTM. 2713T4.

Figure 414. Wagon jack. Cast iron and steel, wood. DTM. 22512T3.

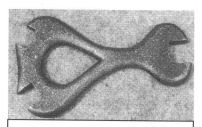

Figure 415. Wagon wrench. Drop-forged iron. CLT. 42912LTC3.

Figure 416. Adjustable wagon wrench. Drop-forged steel, wooden handle, brass ferrule. DTM. 41801T5.

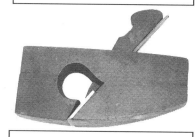

Figure 417. Carriage-makers' plane. Wood (maple), cast steel. CLT. 122712LTC3.

Figure 418. Wagon wrench. Drop-forged steel, brass spring. Reed & Co. of Higgaum, Connecticut. DTM. 21912T1.

Figure 419. Carriage-makers' plane. Wood with steel blade. DTM. 51703T1.

Figure 420. Double jaw buggy wrench. Drop-forged steel, wooden handle. Portland Wrench Co., H.A. Thompson patent. DTM. 041403T1.

Figure 421. Wagon wheel measuring tool or traveler. Forged iron, brass ferrule, wooden handle. DTM. 7602T4.

Figure 422. Wagon wheel measuring tool or traveler. Malleable iron, wood. Wiley & Russell Mfg. Co. DTM. 32708T44.

Figure 423. Carriage-makers' body knife. Cast steel, brass ferrules, and wood handles. DTM. 30911T5.

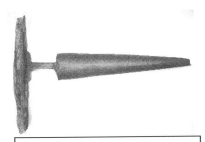

Figure 424. Hub reamer auger. Forged iron, wood. DTM. 72013T3.

Patternmaker

Figure 425. Crane neck gouge. Cast steel, brass ferrule, wooden handle. Buck Brothers. DTM. 42904T4A.

Figure 426. Trammel points. Bronze and oak. DTM. 32708T51.

Figure 427. Core box plane. Cast bronze, wood (rosewood), steel cutter. CLT. 81212LTC13.

88

Figure 428. Patternmakers' mold. Wood, black paint. DTM. 2713T2.

Figure 429. Patternmakers' mold. Cast brass. CLT. 21013LTC1.

Figure 430. Patternmakers' molding tool. Forged iron, wood (boxwood). DTM. 21013T2.

Figure 431. Pin vise. Cast steel. DTM. 102100T9.

Figure 432. Patternmakers' tools. Cast steel, brass, and bronze. DTM. TCT1004, TCT1301, 62202T9, TCT1006, TCT1007, 121600T3, TCT1003.

Figure 433. Patternmakers' molding tools. Drop-forged steel. DTM. 6113T2.

Figure 434. Patternmakers' block plane. Wood (boxwood), cast steel. DTM. 62202T5.

Figure 435. Patternmakers' block plane. Cast steel, wood (lignum vitae). DTM. 100400T3.

Other Early American Trades

Cobbler

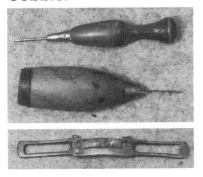
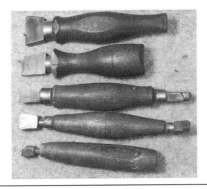

Figure 436. From the cobblers' chest on the 2nd floor of the Davistown Museum. Above, left to right: Awls, heel shave, creasing irons, hammers.

Figure 438. Cobblers' channel gouge. Drop-forged steel, wooden handle, brass ferrule. DTM. 3312T3.

Figure 439. Cobblers' corrugated burnisher. Steel, brass ferrule, wood (beech). DTM. TCH1005.

Figure 437. Cobblers' burnisher. Forged iron, wooden handle. DTM. 102904T9.

Figure 440. Cobblers' hammer. Drop-forged steel, wooden handle. DTM. 22411T2.

Figure 441. Cobblers' hammer. Drop-forged steel, hickory wood handle. DTM. 32412T4.

Figure 442. Revolving leather hole punch. Steel. W. Schollhorn of New Haven, Connecticut. CLT. 42812LTC2.

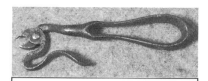

Figure 443. Cobblers' tack puller. Forged or cast iron. DTM. TCH1002.

Figure 444. Leather chamfer. Wood, brass, steel. DTM. 111001T35.

Figure 445. Draw gauge. Drop-forged steel and iron, brass, rosewood. C. S. Osborne & Co. DTM. 102904T20.

Figure 446. Round leather hole punch. Cast steel. DTM. 8912T5.

Figure 447. Rawhide mallet. Rawhide, wood (hickory). Garland. CLT. 92812LTC1.

Figure 448. Leather washer cutter. Drop-forged steel. CLT. 9912LTC3.

Figure 449. Cobblers' toe or beating out hammer. Cast steel, wood (hickory). United Shoe Machine Company. CLT. 93012LTC5.

Figure 450. Head knife. Drop-forged steel, brass, wooden (rosewood) handle, leather sheath. C. S. Osborne. CLT. 10112LTC3.

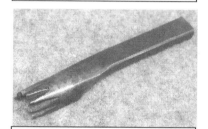

Figure 451. Pinking iron. Cast steel. C. S. Osborne. CLT. 101412LTC2.

Figure 452. Leatherworkers' smoothing tool. Rosewood. DTM. 102512T16.

Figure 453. Cobblers' hammer. Steel, wooden handle. Clarkson. DTM. 102512T1.

Figure 454. Cobblers' beating out hammer. Cast steel, wood (hickory). CLT. 3213LTC1.

Figure 455. Cobblers' hammer. Cast steel, wood (hickory). CLT. 22813LTC7.

Figure 456. Shoe or saw toothed tack puller. Steel, Bakelite. United Shoe Machine Company. CLT. 52612LTC2.

Agricultural Tools

Figure 457. Dibble. Drop-forged steel with red painted wooden handle. DTM. 31212T17.

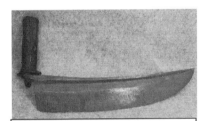

Figure 458. Hay knife. Forged iron and steel, wooden (oak) handle. DTM. 72312T1.

Figure 459. Grain sickle. Steel, rosewood handle. CLT. 42912LTC8.

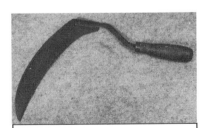

Figure 460. Grass hook or grass sickle. Steel, rosewood. CLT. 42912LTC9.

Figure 461. Hay thief. Forged iron, wood handle. DTM. 102904T16.

Figure 462. Scythe anvil. Wrought iron with a low carbon steel anvil top. DTM. 111406T3.

Figure 463. Veterinary fleam or flem. Steel, brass. DTM. 42712LTC4.

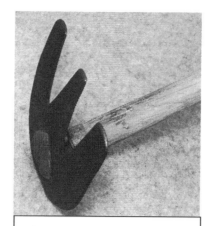

Figure 464. Hoe, forged iron. DTM. 101312T5.

Figure 465. Grub hoe. Forged iron. DTM. 61612T3.

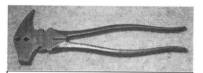

Figure 466. Fencing combination tool. Forged steel. CLT. 42712LTC1.

Figure 467. Grafting iron. Drop-forged malleable iron. DTM. 072112T10.

Figure 468. Grass shears. Drop-forged steel. CLT. 6712LTC3.

Figure 469. Brush hook. Forged steel, wood (hickory). CLT. 33013LTC2.

Figure 470. Pruning knife. Drop-forged steel, brass fittings, wooden (rosewood) handles. Kutmaster.CLT. 72312LTC2.

Figure 471. Dandelion or asparagus knife. Drop-forged steel, wooden (ash) handle. CLT. 7712LTC5.

Figure 472. Pruning shears. Cast steel. DTM. 71312T4.

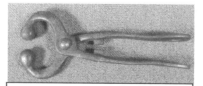

Figure 473. Bull leader. Drop-forged steel. CLT. 8512LTC1.

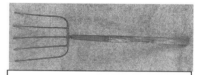

Figure 474. Hay fork. Cast steel, wood (hickory). DTM. 81812T2.

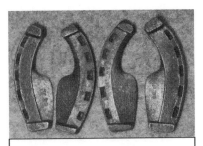

Figure 475. Oxen shoes. Cast steel. DTM. 101312T31.

Figure 476. Grub hoe. Forged steel, wood (hickory). DTM. 22813LTC1.

Figure 477. Butchers' saw. Drop-forged iron and saw steel, wooden handle. DTM. 72712LTC1.

Figure 478. Fence post maul. Wood (oak?). DTM. 102012T1.

Figure 479. Brush cutter. Steel, wooden handle. DTM. 102612T5.

Figure 480. Turf ax. Steel, wood. Joseph Breck & Co. of Boston, Massachusetts. DTM. 102612T3.

Figure 481. Bee smoker. Sheet steel, wood, leather. CLT. 121312LTC1.

Figure 482. Chisel edge pruning saw. Cast iron, malleable iron, and steel. DTM. 7309T4.

Figure 483. Sod cutter shovel. Forged iron, wood (hickory). DTM. 71513T1.

Ice Harvesting

Figure 484. Ice ax. Iron and weld steel. Dernell & Co. of Athens, New York. DTM. 040904T4.

Figure 485. Ice chopper. Drop-forged steel, brass, rosewood handle. CLT. 61512LTC1.

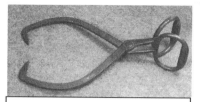

Figure 486. Ice tongs. Forged iron. DTM. 101400T14.

Figure 487. Pitchfork. Forged steel, wood (hickory). DTM. 32313T3.

Figure 488. Ice saw. Steel with cast iron and wood handle. DTM. 71903T2.

Figure 489. Ice chopper. Drop-forged steel, wood (rosewood). Hamilton Beach. DTM. 101312T22.

Maritime Use

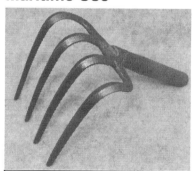

Figure 490. Clam rake. Steel, wood (hickory). DTM. 101312T1.

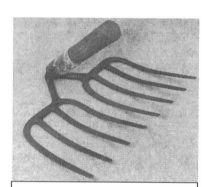

Figure 491. Clam rake. Malleable iron, wood (maple). DTM. 101312T2.

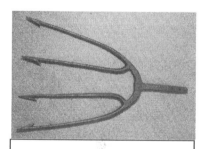

Figure 492. Eel spear. Forged iron. DTM. 102005T1.

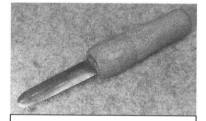

Figure 493. Oyster knife. Steel, wood (beech). CLT. 42812LTC3.

Figure 494. Eel spear. Forged iron. DTM. 22512T8.

Figure 495. Cargo hook. Malleable iron and steel, wood (rosewood). Snow & Nealley Co. DTM. 121412T3.

95

Figure 496. Mackerel plow. Iron, slate, lead, brass, steel, and wood. DTM. 21201T8.

Figure 497. Parallel Rule. Brass, ebony. DTM. 91303T14.

Figure 498. Compass. Cast steel. William Friedricks. DTM. 81101T17.

Domestic Tools

Figure 499. Candle mold. Tin. CLT. 3912LTC3.

Figure 500. Flax hatchel. Wood (maple?), iron. DTM. TAB1013.

Figure 501. Ink eraser. Forged steel, bone. DTM. 42712LTC2.

Figure 502. Embroidery hoop clamp. Iron and steel, japanned finish. Courtesy of George Short. 6712LTC1.

Figure 503. Butchers' cleaver. Forged steel. Foster. CLT. 22813LTC5.

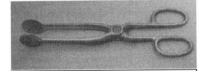

Figure 504. Coal tongs. Forged iron. DTM. 72712LTC10.

Figure 505. Weavers' burling iron. Forged steel. Joseph Lingard. DTM. 52512T1.

Figure 506. Block knife. Forged iron and wood. DTM. 11301T1.

Figure 507. Stove burner handle. Grey cast iron, japanned finish. CLT. 6312LTC3.

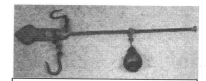

Figure 508. Steelyard scales. Cast iron. CLT. 61512LTC2.

Figure 509. Cigar box opener. Drop-forged steel. R.G. Sullivan. DTM. 42712T1.

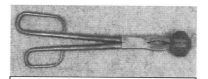

Figure 510. Pinching iron. Drop-forged iron. DTM. 71401T16.

Figure 511. Button hole cutter. Cast steel. W.N. Seymour. DTM. TCP10040.

Figure 512. Peel. Forged iron. DTM. 101400T7.

Figure 513. Cheese sampler. Drop-forged steel or iron. DTM. 121805T24.

Figure 514. Butter knife. Stag horn, shear steel. DTM. 82512T2.

Figure 515. Door latch. Forged iron, wood. Treat of England. DTM. 52603T4.

Figure 516. Wall scraper. Cast iron, steel. DTM. 121412T15.

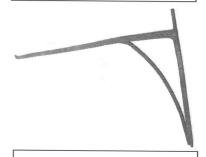

Figure 517. Kettle crane. Forged iron. CLT. 22813LTC6.

Figure 518. Wig blower bellows. Wood, leather, and tin. DTM. 12900T10.

Figure 519. Button wire pliers. Drop-forged steel. CLT. 42912LTC11.

Figure 520. Weavers' shears. Cast steel. Treat of England. DTM. 71312T3.

Figure 521. Ice chopper and scoop. Drop-forged steel, wood (rosewood). North Brothers. DTM. 101312T13.

Figure 522. Glass cutter. Drop-forged steel. DTM. 81713T15.

Currier

Figure 523. Curriers' fleshing knife. Forged iron and steel, brass ferrule, wood handle. D. Tomlinson. DTM. 62406T2.

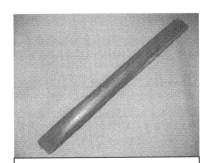

Figure 524. Leather burnisher. Lignum vitae. DTM. 041505T40.

Figure 525. Leather burnisher. Steel, brass ferrules, and wood handles. DTM. 30311T9.

Figure 526. Curriers' fleshing knife. Drop-forged steel, rosewood handles. Snow & Nealley. DTM. 61612T8.

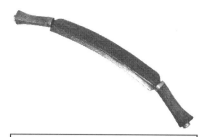

Figure 527. Curriers' fleshing knife. Cast steel, wood (rosewood). Snow & Nealley. DTM. 121412T13.

Sailmaker

Figure 528. Bodkin. Rosewood, cast steel, brass ferrule. DTM. 31112T3.

Figure 529. Fid. Tropical hardwood. DTM. 31112T2.

Figure 530. Marlin spike. Drop-forged malleable iron or steel. DTM. 30911T7.

Figure 531. Cloth knife. Drop-forged steel blade, brass, wooden (rosewood) handle. CLT. 72712LTC5.

Figure 532. Fid. Hardwood (beech). CLT. 81212LTC5.

Figure 533. Fid. Hardwood (apple). CLT. 81212LTC3.

Figure 534. Bodkin. Cast steel, brass ferrule, wood handle (mahogany), ivory inset. CLT. 81212LTC4.

Figure 535. Walrus tusk fid. CRF.

Figure 536. Marlin spike. Cast steel, brass ferrule, hardwood handle (ebony?), ivory butt. CLT. 81212T1.

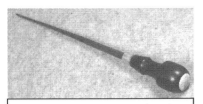

Figure 537. Bodkin. Wood (ebony), forged steel, brass ferrule. CLT. 81212LTC15.

Figure 538. Marlin spike. Forged steel. B.D. Wiley. CLT. 81212LTC7.

Figure 539. Sailmakers' kit. Cast steel, wood, leather, paper, cotton twine. DTM. 42012T6.

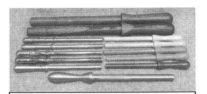

Figure 540. Belaying pin set. Wood (lignum vitae, teak, beech), turned copper, turned brass, cast brass, nickel-plated cast iron. DTM. 9912T2.

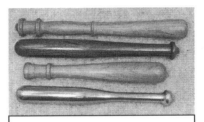

Figure 541. Keeper pin set. Cast brass, wood (maple, redwood). DTM. 9912T3.

Figure 542. Marlin Spike. Drop-forged steel, brass, wood (tiger maple). DTM. 71312T2.

Rope-maker

Figure 543. Cordage rule. Boxwood and brass. John A. Roebling. DTM. 42602T2.

Figure 544. Cordage rule. Boxwood and brass. John A. Roebling. DTM. 82709T1.

Part III: Hand Tools of the Industrial Revolution

The Industrial Revolution: The First 100 Years

The Industrial Revolution didn't suddenly begin when the Samuel Collins Axe Factory in Connecticut (est. 1827) began using machinery to drop-forge axes in 1837. Industrial history in America, and especially in Europe, is characterized by numerous innovations and inventions that paved the way for the mass production of tools and machinery with interchangeable parts, a story told in detail in the first three volumes of the *Hand Tools in History* publication series. This introductory essay is a short summary of that labyrinth of technological rabbit holes.

It may be argued that a proto-Industrial Revolution began with the introduction of the blast furnace in Europe in the 13th century. A wide variety of steelmaking strategies, including the introduction of the cementation furnace in England in the 17th century to make blister steel, characterized the proto-industrial revolution of the Renaissance and the Enlightenment.

Most of the technological innovations of the early Industrial Revolution (after 1742) have roots in the Enlightenment. The rationalism of the French Enlightenment produced Diderot's ([1751-65] 1964-6) great *Encyclopedia* of the iconography of French trades, much of which were devoted to the maintenance of an elite French aristocracy that was overthrown in the French Revolution. It also produced the first attempts to make guns with interchangeable parts, an event that occurred at an armory in Paris where the first such specimens were part of a display in the late 1770s.

The English Enlightenment, much more empirical and pragmatic, began producing technological innovations in the early 18th century, the most important of which was Abraham Darby's use of coke instead of coal to fuel blast furnaces. It can be argued that the first stage of the full-fledged Industrial Revolution began with Benjamin Huntsman's rediscovery of how to forge crucible steel (a form of Wootz steel) for his precision watch components. Called cast steel, Huntsman's pure steel was characterized by evenly distributed carbon content due to the high temperatures achieved in his Stourbridge clay crucibles. Cast steel was quickly adapted for use in precision navigation equipment and "cast steel" edge tools. These hand-forge-welded tools were one of the most important components of the toolkits of early American woodworkers in the last four decades of the 18th century, where their importation from England played a key role in the florescence of New England's extraordinary case furniture industry.

Our arbitrary date of 1827 marks the end of our early industrial period with the establishment of the Collins Axe Factory where drop-forging machinery was introduced by 1837, but as with many other trades and tools, handmade forge-welded tool production continued into the late 19th century. Many of America's gorgeous and well wrought cast steel edge tools (Witherby, Buck Brothers, Underhill clan etc.) were produced between 1850 and 1890 and involved the adaptation of drop-forging to the production of cast steel edge tools, which still required extensive hand work and careful tempering and annealing.

The first extensive production of cast steel in America, not to be confused with modern cast steel produced by the ton in modern furnaces, began after Joseph Dixon perfected the use of the graphite crucible to make cast steel. By the late 1850s, American production of domestic cast steel equaled the quality of English cast steel. Much of the gradual replacement of water-powered and charcoal-fueled forging and toolmaking processes by steam-powered machinery, often utilizing coal and coke as the fuel, occurred during this period. The florescence of a classic period of American cast steel edge tool manufacturing was well-established by the 6th decade of the 19th century. When shipbuilding began declining after 1860, the demand for edge tools for the shipwright was gradually replaced by the need for high quality durable patternmakers' tools for the vast industrial production of machinery after 1850, all of which needed wooden patterns for their casting. The earlier use of patternmaking tools for making wooden molds for cast iron production (e.g. pots, kettles, hardware, etc.) provided the basis for the expansion of the tool kits of the patternmaker to meet the challenges of a rapidly expanding Iron Age of machinery production.

Many cast steel edge tools used by the shipwright, patternmaker, and timber framer made before the era of Bessemer's bulk processed steel production still make their appearance today in New England tool chests and collections, though production of cast steel edge tools ended with the demise of both charcoal fired wrought iron and crucible steel production in the third decade of the 20th century.

At this time, millions of tons of low carbon cast steel were being produced to build skyscrapers – an entirely different critter from the high quality cast steel formerly produced only in crucibles. Modern steelmaking technologies also resulted in the perfection of the production of tempered alloy and stainless steels, resulting in the mass production of steel and grey cast iron tools cast in molds by the drop-forging process (Figure 545). The skills of ancient ironmongers and their forge-welded tools became irrelevant as science unraveled the secrets of the microstructure of steel. Edgar Bain's (1939) *Functions of the Alloying Elements in Steel* marked the beginning of an age of hi-tech alloy steel production, much of it as cast steels. In the era of information and communication technology, nanotechnology now produces silicon chips, an offshoot of our expanding knowledge of the microstructure of steel.

For definitions about steelmaking technologies and the chemistry of iron and steel, see the Davistown Museum's *Handbook for Ironmongers* (Brack 2013). The recent second edition now includes contemporary definitions of carbon, alloy, and high speed tool steels.

The English Industrial Revolutionaries

The invention and production of cast steel is only one chapter in the history of the first century of the mature Industrial Revolution. The most important component of this revolution was the invention and perfection of the steam engine. This prime mover of industrial society had its roots in the invention of the Newcomb engine with its pistons and cylinders, circa 1712, to pump water out of coal mines to facilitate blast furnace operations and home heating in England. Between 1763 and 1769, James Watt redesigned and improved this engine with the help of John Wilkinson's engine cylinder (invented in 1758). Luckily, for American

Figure 545. Set of patternmakers' dies for surgical tools. Steel. DTM. 8213T1.

colonists and the success of the American Revolution, Matthew Boulton and James Watt didn't begin their mass production of the steam engine until 1775. Its earlier production would have given the English time to establish a more efficient proto-factory system for producing cannons, guns, and ammunition, possibly changing the outcome of the American Revolution.

The period between 1775 and 1825 is an amazing half century of new inventions and innovations, mostly by British, rather than by American or French, inventors. Among the most important of all innovations was Henry Cort's redesign of the reverbatory furnace to produce malleable iron from cast iron without fuel contact and the absorption of sulfur that resulted from this interaction (1784). In the same year Cort invented and patented grooved rolling mills for the production of large quantities of malleable iron-derived bar stock and sheet iron for the expanding economies of Britain and America. The reverbatory furnace and associated rolling mills soon appeared in America, producing larger quantities of iron, often referred to as puddled iron, than could be manufactured in the now antiquated rolling mills of the finery and chafery. The American version of the Industrial Revolution was soon off and running.

While the invention and perfection of the steam engine, reverbatory furnace and grooved rolling mills was occurring in England, other English Industrial Revolutionaries were preparing the way for the mechanization of textile equipment, which revolutionized textile production in England in the late 18th century. The evolution of a mechanized textile industry was the result of a series of English inventions: the flying shuttle, 1738, was followed by the spinning jenny, 1764, which was improved for warp weaving in 1769. Water-powered looms, first introduced in 1771, were improved for carding and roving by 1775. Samuel Slater soon moved to America and established the first mechanized textile mill in Pawtucket, RI, in 1795, and New England's growing population soon had cheap, mass produced clothing. Early American domestic industries, such as weaving on hand looms began declining. The first fully automated textile mill was constructed at Waltham, MA, in 1813.

A now forgotten, but most important series of innovations, were made by Henry Maudslay in England between 1802 and 1807. Maudslay designed and produced 45 different types of machines for making the blocks, as in block and tackle, for the sailing ships of the British navy, which was rapidly conquering large areas of other continents to compensate for its unfortunate loss of the American colonies. Many of these machines were adopted for use by American toolmakers after 1820, who were adept at pilfering or copying English designs, which were often not adapted for use by hidebound English manufacturers. For sociological reasons, the English had difficulty in applying many of their own inventions for practical use in comparison with the more innovative and proactive American entrepreneurs. A more detailed review of the many inventions of the English industrial revolutionaries is found in *Steel- and Toolmaking Strategies and Techniques* (Brack 2008, 83-7).

The Early Years of Prosperity and Change

In America, the adaptation of the innovations of the English Industrial Revolution didn't occur until after the great prosperity of the neutral trades (1789-1807), symbolized by the construction of the many gorgeous federal period homes in New England coastal communities prior to the end of the War of 1812. America began adopting those obscure French innovations in gun production with interchangeable parts, led by Maine's John Hall at the federal forge in Harper's Ferry, VA. Eli Terry introduced interchangeable wooden, and then brass and iron components in his Connecticut clock-making factory well before Samuel Collins began drop-forging his axes. Thomas Blanchard designed the first lathe for turning gunstocks in 1818 based, in part, on designs by those hidebound English industrial revolutionaries. Eli Whitney attempted to introduce gun production with interchangeable parts in his armory during the second decade of the 19th century, an effort which was not perfected until shortly after his death in 1825. Using the Englishman

Samuel Lucas's 1804 advances in the art of making cast iron malleable, Seth Boyden began commercial production of malleable cast iron in 1831 in New Jersey. At the same time rapid advances were made in the production of a variety of types of grey cast iron from annealed white cast iron. The rapid changes that occurred in America's expanding ability to make hand tools, clocks, and guns culminated in the use of drop-forging by the Collins Axe Factory, an event that marks the beginning of a full-fledged Industrial Revolution in America.

Figure 546. Ax head castings. Cast steel. DTM. 102612T16.

Figure 547. Ball peen hammer. Forged iron. DTM. TCR1002.

By 1840, numerous innovations by both English and American technicians paved the way for the rapid expansion of the American system of mass production of guns, machinery, and tools using interchangeable parts. The reverbatory furnace had been adapted for the production of puddled steel, higher in carbon content than the wrought and malleable iron produced with the help of the rotary puddle ball squeezer, as in "puddled" iron. Use of the hot blast to increase furnace efficiency and output, adopted in England in 1828, along with Joseph Nasmyth's steam powered rotary engine (1837), soon were used in both American blast and bloomery furnaces, greatly expanding the availability of malleable (puddled) iron. Reverbatory furnace and bloomery produced iron could now be easily made with a wide range of carbon content ranging from pure wrought iron (< 0.08 cc) to malleable iron for toolmaking with a carbon content ranging up to that of steel (> 0.5 cc).

The Coming of the Steam Hammer

In 1842, the English industrial revolutionary Joseph Nasmyth also invented the steam hammer, which was essentially a marriage of the steam engine, less than a youthful century old, to that ancient simple machine, the trip hammer. This union resulted in the mass production of the most important prime mover of the Industrial revolution (apologies to textile machinery and gunsmithing enthusiasts,) the railroad engine. The steam engine and the steam hammer have numerous important cousins, aunts, and uncles. If such a thing as a modern Iron Age (read 19th century) can be postulated to be the predecessor of the age of steel (read Bessemer and open hearth steel production > 1870), a most important relative was the cupola furnace, one of the essential improvements of

19th century industrial production. The progeny of the blast furnace, numerous subtle differences in the design of the cupola furnace allowed founders to reprocess pig iron from the blast furnace into every conceivable type of cast iron product. The increasing use of the steam hammer in foundries and forges coincided with the development of larger cupola furnaces, which produced an increasing variety of cast iron products including grey cast iron and malleable cast iron. Malleable cast iron soon included a variety of forms of annealed malleable iron, many of the formulas of which were secret and now lost. Soft grey cast iron joined "semi-steels" made from grey cast iron, which underwent further heat treatments. All had two important characteristics in common: they contained significant amounts of silicon ranging from 0.7% to 2.5%, and they were highly machinable.

These advances stimulated the production of cast iron implements, tools, and machinery of every description, including those still famous icons of the woodworking trades, Bailey then Stanley grey cast iron planes. The production of American-made cast iron cooking pans continued unabated in the late 19th century even as bulk processed steel rails, structural steel, and reinforcing steel production surpassed the productivity of the cupola furnace in the last quarter of the 19th century. For the half century before the perfection of bulk steel production, the rapidly growing need for steel was filled by the continued production of German steel (decarburized cast iron), the perfection of puddled steel production in the refractory furnace, and the continued production of blister steel. Crucible "cast steel" constituted only a tiny percentage of steel production throughout the 19th century, always for specialized purposes, such as edge tool production , razor blades, watch springs, and other precision equipment. Cast steel was gradually replaced for many uses by the introduction of modern tempered alloy steels after the American Civil War, though these steels were often cast, not in crucibles for smelting, but in molds for mass production of railroad and agricultural equipment, and hand tools of every description.

By 1851, Joseph Whitworth, another English industrial revolutionary, perfected the art of constructing industrial machinery, designing and manufacturing his famous screw-cutting lathe to produce the standard screw-cutting threads he had previously invented in 1841. His sophisticated shaping, slotting, drilling, milling, and planing machines and his invention or at least adaptation of a casting process for manufacturing durable, ductile steel are now forgotten enablers of the American factory system. Did his use of ductile steel play a role in Leonard Bailey's pioneering production of durable metal hand planes soon to replace the ubiquitous, if ancient, wooden designs of smoothing and joining planes? The rapid advances in making high quality malleable iron, ductile steel, malleable cast iron, and grey cast irons were well kept secrets of the ironmongers who flourished using empirical rule-of-thumb procedures. These secrets would soon be unlocked by the science of chemistry and the growing knowledge of the role of carbon in ferrous metallurgy. Ever larger, smoke-belching (hello CO_2 and methylmercury) factories began making machinery of all kinds. Stashed away in obscure workshops was a growing repertoire of hand tools whose production was facilitated by the advances in making iron and steel that made the factory system of mass production possible. Eventually, the hand tools used to build machinery, built machinery to make hand tools. A virtual explosion of the production of hand tools of every description, but especially those of the machinist and mechanic, coincided with the rapid spread of the use of alloy steels, including tempered alloy steels and high speed tool steels. The invention and use of the micrometer and vernier calipers, soon to have a vast multiplicity of forms, quickly followed. In the early 20th century, Harry Brearley's perfection of the art of making stainless steel (1913) was a milestone in the classic period of American toolmaking, which soon entered a period of gradual decline.

American Cast Steel and the End of the Iron Age

In 1850, the American Joseph Dixon had invented the high temperature resistant graphite crucible, which made mass production of American-made cast steel edge tools possible. Good-bye shipbuilding tools (well, not quite yet), hello patternmakers' tools and Mr. Thomas Witherby, 1849, and John, Charles, and Richard T. Buck, 1853.

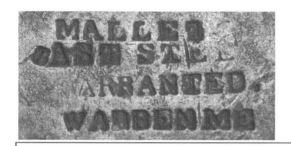 

Figure 548. Maker's marks from socket chisels, American on left, English on right. Cast steel. DTM. 102904T12 and 100108T3.

Two portents of the end of the Iron Age occurred in 1856, the obscure invention of gasoline in Watertown, MA, and Henry Bessemer's famous announcement of a new technique of making malleable iron (low-carbon steel) without fuel. The latter announcement preceded an innovation that equaled in significance the invention, development, and use of the steam hammer. Robert Mushet, yet another of the English industrial revolutionaries, invented and produced the first alloy steel in the forest of Dean in England 1862-1866. Perfected by 1868 as "self-hard" air cooled steel (8% tungsten, 2% carbon, 1% manganese), this steel alloy reigned for 30 years as the primary component of machinists' cutting tools used on the lathes and cutting machines of America's now rapidly expanding factory system.

A fundamental characteristic of the American factory system was the manufacturing of tools with interchangeable parts. Those one-of-a-kind individually made machinists' tools or tool forms made only in small quantities, soon became tool forms mass produced in ever increasing numbers. Unique tool forms for a wide variety of trades began appearing as the American factory system gave birth to a proto-consumer society. Domestic and urban environments as well as transportation and industrial infrastructure required maintenance utilizing hand tools of every description.

Early American Industrial Transition

The techniques used for tool production by most early American industries didn't suddenly come to a halt between 1827 and 1837 when Samuel Collins adapted some of the English machinery designs and introduced the drop-forging technique using dies for making axes. Many techniques, technologies, and trades of early American industries lingered on into the late 19th century. In Maine, careful hand work at the ax manufacturers on the tributaries of the Kennebec River continued until, in some cases, the final closing of some factories in the 1970s. Many a coopers' adz, mast ax, or drawknife made before 1837 looks indistinguishable from those made after the Civil War. Yet unique tool forms, some used to make the machinery that made handmade tools obsolete, began to appear in the early 19th century tool chests and workshops of New England artisans. Their use and iconography signal a coming transition. Initially, handmade hand tools built the machinery of a factory system that soon made tools with interchangeable parts or with identical forms. The mass production of hand tools soon followed. An iconography of the tools characterizing the florescence of the classic period of the American factory system follows these introductory essays.

The Stanley Tool Company: A Paradigm

One company, the Stanley Tool Company, having two major components with ever changing names, best illustrates the birth and growth of the American factory system and its classic period of American toolmaking.

Augustus Stanley began making rules in New Britain, CT, possibly as early as 1850. Another Stanley family company, The Stanley Works, maker of hand tools of all description other than rules and levels, was organized in 1856 (Jacob 2011, 6). In 1858, the A. Stanley Rule Co. merged with Hall & Knapp, makers of try squares, levels, and plumbs, and became the Stanley Rule & Level Company. Both Stanley companies operated simultaneously until they merged in 1920. The history of their hand tool manufacturing output, as illustrated by the myriad variety of hand tools they produced, is a paradigm of the evolution of America's factory system of hand tool production.

108

Members of the Stanley clan had been making hinges and hardware in New Britain since 1831. The ease of production of high quality wrought iron in refractory furnaces for the last half century or longer made machine-made, rather than hand forge-welded, hardware production a routine affair. When the Stanley Works was organized in 1856, machinable grey cast iron (±2.75% carbon content) was available for tool production as was high carbon malleable cast iron (±3.4% carbon content) and tool steel (> 0.5% carbon content). Soon to make an appearance for hardware and tool production was cold rolled Swedish iron, first used by the Stanley Works in 1870 (Jacob 2011, 9) but with a long history of importation to the United States for edge tool making and other uses. Puddled steel had been available for drop-forging tools for several decades; the appearance of tempered alloy steel with various degrees of hardness evolved in the late 1860s. The availability of higher carbon grey cast iron and malleable cast iron with a slightly lower carbon content expanded tool production options. Malleable iron was soon to be called mild steel and used in many toolmaking operations. Numerous variations in hot or cold rolling, tempering, and annealing, many of them secret, were adopted by the Stanley Works and soon by Darling, Brown, & Sharp, the L.S. Starrett Company, and other American toolmakers working during the classic period of American toolmaking. Other than its excellent Sheffield cast steel edge tools and numerous high quality planemakers, English toolmakers were falling behind in their production capabilities compared to the innovative entrepreneurs of the American factory system. The hand tools that built the machinery that mass produced other hand tools and manufactured other machinery are a starting point in the exploration of the iconography of factory-made American hand tools. The Stanley legacy of high quality factory-made hand tools is one important chapter in the narration of this story.

The Classic Period of American Toolmaking

By 1875, the heyday of the classic period of American toolmaking was underway. As a result of the proliferation of new tool- and steelmaking strategies and techniques, which also included the Siemens open hearth furnace (1863), a robust community of New England toolmakers, including machinist toolmakers, soon evolved in response to the rapidly increasing demand for hand tools for many specialized trades. Martin Donnelly's (1993) list of *Classic American Machinist Toolmakers* delineates the principal machinist tool manufacturers who formed the core of the heyday of the classic period of American toolmaking (listings in bold are on Donnelly's original list).

Maker	Location	Working Dates	Products
Athol Machine Company	Athol, Mass.	1868-1920	calipers, dividers, squares, etc.
Bemis & Call H. & T. Co.	Springfield, Mass.	1844-1910	calipers, dividers & squares, etc.

Maker	Location	Working Dates	Products
Billings & Spencer Co.	Hartford, Conn.	1882-1914	gauges, measuring tools, calipers, etc.
Brown, J.R.	Providence, R.I.	1841-1853	rules, gauges, etc.
Brown & Sharpe, J.R.	Providence, R.I.	1853-1867	rules, gauges, etc.
Brown & Sharpe Mfg. Co.	Providence, R.I.	1868-Present	micrometers, rules, squares, etc.
S.W. Card Mfg. Co.	Mansfield, Mass.	1874-1908	taps, dies, etc.
Coffin & Leighton	Syracuse, New York	1885-1901	rules & gauges
Cook, J.H.	Syracuse, New York	1890-1902	rules, gauges, etc.
Darling & Bailey	Bangor, Me.	1852-1853	rules
Darling, Brown & Sharpe	Providence, R.I.	1866-1895	rules, gauges, squares, etc.
D. & S. (Darling & Schwartz)	Bangor, Me.	1853-1866	rules, squares, etc.
Davis, L.L.	Springfield, Mass.	1867-75	levels
Davis Level & Tool, Co.	Springfield, Mass.	1875-1893	levels
Fay, Charles	Springfield, Mass.	1883-1887	calipers & dividers
J. M. King & Co.	Waterford, NY	1887-1910	dies, pliers & taps
Massachusetts Tool Co.	Greenfield, Mass.	1900-1925	squares, gauges, rules, etc.
Sawyer Tool Company	Athol, Mass. Fitchburg, Mass. Ashburnham, Mass.	1894-98 1898-1912 1912-15	squares, rules, gauges, etc.
Standard Tool Company	Athol, Mass.	1882-1905	squares, calipers, gauges, levels, etc.
Starrett Co., The L.S.S.	Athol, Mass.	1880-Present	squares, calipers, gauges, etc.
J. Stevens & Company	Chicopee Falls, Mass.	1864-1903	calipers, gauges, etc.
Stevens Arms & Tool Co.	Chicopee Falls, Mass.	1886-1903	guns, bits, calipers, dividers, levels
Union Tool Co.	Orange, Mass.	1908-75	rules, squares, gauges, etc.
Walker Co., Edwin	Erie, Pa.	1887-90	surface gauges & marking gauges
John Wyke & Co.	E. Boston, Mass.	1885-1911	gauges, rules, etc.

Accompanying the rise of the machinist with his increasingly complex toolkits were newly evolving specialized trades, such as plumbing and surveying, soon followed by electricians and the proliferation of small workshops for producing automobiles. Other long established traditional trades adopted new tool forms, including the silversmith, whitesmith, quarryman, gunsmith, jeweler, and clockmaker. Sewing machine production had already played a role in the expansion of the American factory system, soon to be joined by the typewriter, bicycle, and harvesting equipment production. An amazing variety of tool forms were invented and manufactured, often by small obscure toolmakers, to service the growing repertoire of complex machines. A review of many of the tool forms illustrated in Diderot's ([1751-65] 1965-1966) *Encyclopedia* (see *Appendix 1*) reveals that many of the new tool forms produced by American toolmakers in the 19th century are copies of or adaptations of the amazing variety of hand tools being used by French society and its many artisans in the early and mid-18th century. In turn, some of these tool forms can be traced back to Roman times, although a systemic study of the roots of modern hand tools (read post-1700) in the early Iron Age (1200 BCE to 1200 CE) has never been published.

The Tools that Built Machines

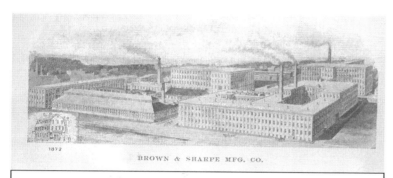

Figure 549. Brown & Sharpe Mfg. Co. From the *Brown & Sharpe Mfg. Co. Catalogue*, 1902.

The hand tools and machinery used in the first workshops and small factories at the beginning of the classic period of the American Industrial Revolution (1827) were not mass produced in factories, in contrast to the vast production of the mammoth factories, such as the Brown & Sharp industrial complex operating by the last decades of the 19th century. Some of the first lathes and proto-milling machines were either one-of-a-kind designs or produced in small numbers with carefully forged or cast parts. By 1865, the availability of a wide variety of malleable iron, puddled steel, grey cast iron, or malleable cast iron gave tool- and machine-makers a much wider choice of iron products to use in manufacturing hand tools than were available before 1840.

Among the most important tools used by machine-makers were calipers, the use of which can be traced back to Bronze Age shipwrights, who made the same measurements as 19th century shipwrights building schooners for the coasting trade. Initially, wood, then hand-forged wrought and malleable iron, and finally threaded adjustable (inside, outside, etc.) calipers made from puddled steel made their appearance.

Figure 550. O. R. Chaplin's patent combination square head. Cast iron, drop-forged steel. DTM. 92212LTC1.

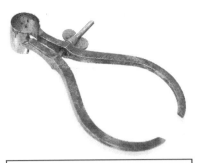

Figure 551. Fay calipers. Cast and forged steel. Signed "P.S. Stubs". CLT. 2713LTC1.

Figure 552. Outside calipers. Forged iron. DTM. 32802T1.

Figure 553. Brass square. CRF.

Figure 554. Davis corner level and inclinometer. Cast iron, brass, glass. CRF.

Figure 555. Hand vise. Drop-forged iron or steel. W. & C. Wynn. DTM. 121112T6.

Figure 556. Precision square. Tool steel. Taft & Pierce Mfg. Co. DTM. 81713T14.

As soon as the production of identical copies of the same machinery became a strategy of the factory system of mass production, patterns to mass produce identical components of machinery as well as stationary engines, wheels, and turbines made their appearance. The casting of machine and engine components in sand casts required the initial production of

112

wooden patterns of the components of machinery to be cast. These wooden patterns were handmade using patternmakers' tools, the proliferation of which signaled the beginning of the factory system of mass production. A selection of patternmaking tools are illustrated in Figure 425 to Figure 432. Edge tools not too dissimilar to those illustrated in Figure 21 (Moxon [1703] 1989) were still used to shape the wood patterns that were used to make the sand molds into which cast iron, including malleable cast iron, would be poured for machine components. Molding tools and smoothers (Figure 101), not to mention specialized tools such as the core box plane (Figure 427), signaled a whole new era of industrial expansion that was dependent upon obscure hand tools that built machinery. Other tools characteristic of early machine shops include depth gauges, surface gauges, height gauges, drills, drill indexes, and screw plates. Many of the hand tools of the multitasking blacksmith also continued to be used by the artisans who built the first machines of the early Industrial Revolution.

Machinists gradually replaced woodworkers as prime movers of what also might be called the classic period of America's Industrial Revolution. The machinist, along with chemists and inventors such as Thomas Edison, in turn became the prime movers of what can be called the age of petrochemical-electrical man. The iconography of hand tools narrates the history of our industrial society, now entering the age of fiber optics-driven information technology. Manganese may still be incorporated for edge tool production, which still continues on a small scale; it also often appears in alloy steel and tool steel production, and is always a component of modern carbon steels. It is silicon, now the most famous of non-metals, which has re-emerged as the key component of our hyper-digital flat world of information and communication technology.

Machines as Prime Movers

The first lathes, drill presses, and milling machines weren't the only prime movers of the Industrial Revolution. Steam engines = steam boats was an equation universally recognized by 1820. The construction of the first railroad between Boston and Worcester in 1835 is another key event in that decade of rapid industrial change just before drop-forging technology made widespread production of machinery and tools with interchangeable parts possible via the American factory system. The rapid speed of the expansion of industrial productivity in America left the English industrial revolutionaries puzzled – why were their innovative machinery designs being so rapidly adapted in the mostly still unexplored continent of open horizons?

By 1840, American production of steam engines and railroad equipment had surpassed English production and equaled its quality. Machine tools were the instruments of manual operation used to create these Industrial Revolution prime movers. Tools produced to operate and repair this machinery, illustrations of which follow the iconography of machinist tools, is implicit in the cascade of inventions characterizing the second half of the 19th century, including the telegraph, telephone, electric light, automobile engine, bulk processed steel production, tilting band saw, milling machine, and a wide variety of other equipment. Mass production of hand tools, illustrations of which follow those of tools used to operate and repair machinery, are the progeny of this cascading technological change. Railroads, steam engines, and the machinery they powered were the prime movers of a hand tool-producing industrial society until petro-chemical-electrical man formulated the next Industrial Revolution.

Machinists' Tools

Figure 557. Machinists' combination square. Drop-forged tempered alloy steel, glass, satin and japanned finish. Brown & Sharpe. CLT. 22512LTC4.

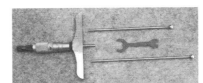

Figure 558. Depth micrometer. Drop-forged tempered alloy steel. Starrett. CLT. 22512LTC9.

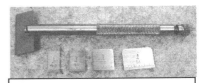

Figure 559. Machinists' scale set. Stainless steel blades, tempered alloy steel handle. Starrett. CLT. 3912LTC2.

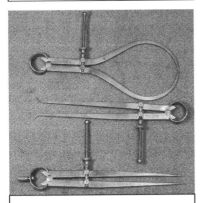

Figure 560. Caliper and divider set. Drop-forged tempered alloy steel. L.S. Starrett. CLT. 22512LTC1.

Figure 561. Small hole gauges. Drop-forged tempered alloy steel, vinyl pouch. L.S. Starrett. CLT. 22512LTC7.

Figure 562. T-handle tap wrench. Drop-forged steel. Dikeman Mfg. Co. of Norwalk, Connecticut. DTM. 22612T9.

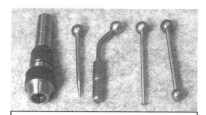

Figure 563. Machinists' wiggler center finder. Drop-forged tempered alloy steel, vinyl pouch. L.S. Starrett. CLT. 22512LTC5.

Figure 564. Ruler clamp. Tempered alloy steel. L.S. Starrett. CLT. 4512LTC1.

Figure 565. Toolmakers' surface gauge. Drop-forged tempered alloy steel. CLT. 22512LTC12.

Figure 566. Center gauge. Tempered alloy steel. L.S. Starrett. CLT. 42912LTC17.

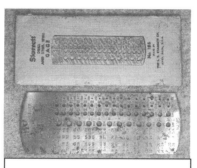

Figure 567. Drill and wire gauge. Tempered alloy steel. L.S. Starrett. CLT. 22512LTC8.

Figure 568. Inside micrometer calipers. Drop-forged tempered alloy steel. CLT. 22712LTC1.

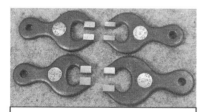

Figure 569. Machinists' go nogo caliper gauge set. Drop-forged steel. DTM. 52512T2.

Figure 570. Metallic plumb and level. Grey cast iron, steel, nickel plating, japanned finish. CLT. 52712LTC3.

Figure 571. Starrett No. 13 Double Square. Tool steel. CLT. 8512T1.

Figure 572. Firm joint hermaphrodite calipers. Tool steel. L.S. Starrett. CLT. 62112LTC1.

Figure 573. Spring dividers. Tempered alloy Steel. L.S. Starrett. DTM. 1302T3.

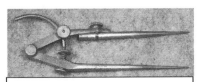

Figure 574. Extension dividers. Tempered alloy steel. L.S. Starrett. DTM. 32708T46.

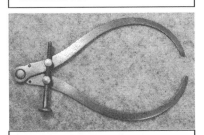

Figure 575. Outside screw adjusted calipers. Forged iron. W.H. Hale. DTM. 83102T8.

Figure 576. Spring winder. Forged steel, brass, cast iron handle, wooden handle. DTM. TJE1003.

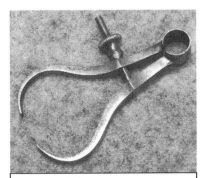

Figure 577. Outside screw adjusting spring calipers. Tempered alloy steel, brass. Joshua Stevens Arms & Tool Co. DTM. 032203T9.

Figure 578. Line level. Drop-forged steel. H.B. Brown. DTM. 102503T3.

Figure 579. Boulet's indicator. Steel. Boulet's Fine Tool Works, Sebago Lakes, Maine. DTM. 102503T1.

Figure 580. Wire gauge. Steel. The Blodgett Mfg. Co. of Rochester, New York. DTM. 10910T4.

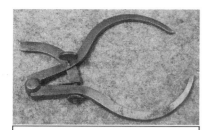

Figure 581. Wing calipers. Forged Steel. Bemis & Call. CLT. 5412LTC7.

Figure 582. Wire gauge. Drop-forged tempered alloy steel. J.R. Brown & Sharpe. CLT. 22412LTC1.

Figure 583. Go nogo inspection gauges. Drop-forged steel, japanned finish. Brown & Sharpe. DTM. 71912LTC1.

Figure 584. Wire gauge. Steel. Morse Twist Drill & Machine Co. of New Bedford, Massachusetts. DTM. 10910T5.

Figure 585. File handle. Cast iron. CLT. 72712LTC4.

Figure 586. Starrett No. 133 Adjustable incline level. Cast steel, japanned finish, glass vials. LTC. 71412LTC1.

Figure 587. Steel plumb bob. Drop-forged steel, nickel-plated finish. L.S. Starrett. LTC. 7712LTC2.

Figure 588. Die stock. Cast steel. Wiley & Russell Mfg. Co. DTM. 33112T1.

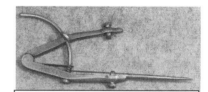

Figure 589. Point dividers. Cast brass, steel. CLT. 81212LTC11.

Figure 590. Toolmakers' buttons. Tool steel. L.S. Starrett. DTM. 8512T1.

Figure 591. Caliper wire gauge. Cast steel. George Partridge of Darlaston, England. DTM. 42012T4.

Figure 592. Square and round machine files. Drop-forged tool steel. Nicholson. DTM. 31212T9.

Figure 593. Jack screw. Drop-forged steel. DTM. 8312T5.

Figure 594. Wing divider. Cast steel. C. Delsten & Sons. DTM. 31908T29.

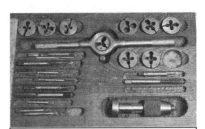

Figure 595. Tap and die set. Drop-forged tempered alloy steel, drop-forged steel bits. CLT. 22512LTC6.

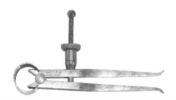

Figure 596. Outside calipers. Tempered alloy steel. L.S. Starrett. DTM. 111412T5.

Figure 597. Dividers. Tempered alloy steel. L.S. Starrett. DTM. 111412T5.

Figure 598. Firm-adjust calipers. Steel. D.E. Lyman. DTM. 102503T2.

Figure 599. Wing divider. Iron and cast steel. Harrington. DTM. 111412T4.

Figure 600. Wing divider. Cast steel. DTM. 31311T9.

Figure 601. Marking gauge. Tool steel. DTM. 040103T5.

Figure 602. Threading tool holder. Drop-forged steel. J.H. Williams. CLT. 32313LTC7.

Figure 603. Threading tool holder. Drop-forged steel. Pratt & Whitney. CLT. 32313LTC8.

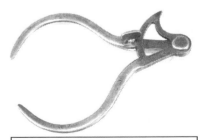

Figure 604. Wing divider. Cast steel. W. Schollhorn. CLT. 32313LTC4.

118

The Tools That Operated and Repaired Machines

Unique machines, some produced in small quantities and others soon mass produced quickly stimulated an innovative tool industry, often of small independent entrepreneurs who began the design and manufacturing of a principal tool characteristic of the modern age, the wrench. Just as edge tools were the typical tools of the Wooden Age, the wrench is a symbol of the arrival of the machine age. Only the implements of the tool and die maker (the machinist), hidden away in the workshops of vast factory complexes, equal the importance of the wrench in the birth, growth, and functioning of the American factory system.

Wrenches

Wrenches are a simple machine that can be traced back to the Bronze Age. Primitive cast iron adjustable wrenches were the most common wrench form from the Roman era to the late 18th century (Figure 4). Adjustable wagon and carriage wrenches, initially made in England but soon manufactured in or near Boston, still make an occasional appearance in New England workshops (Figure 1).

Figure 605. Center adjusted nut wrench. Drop-forged iron, wood (rosewood). Solymon Merrick. DTM. 62406T6.

Figure 606. Adjustable nut wrench. Cast steel, wood (rosewood). Coes. CLT. 52313LTC1.

Figure 607. Coes Wrench Company (Franklin 1899, 503).

The appearance of the "modern" monkey wrench, among the first drop-forged tools, coincides with the proliferation of steam engines, especially as incarnated in the railroad engine. Among the earliest monkey wrenches, and probably a predecessor of the Coes Wrench Company wrenches is the Solymon Merrick adjustable wrench (Figure 605). The Coes Wrench Company of Worcester, MA, was the first large factory mass producing these still commonly encountered simple machines.

Between 1860 and 1920, a large community of often obscure and now forgotten inventors and entrepreneurs created an amazing variety of wrenches, most now having the status of "collectibles," to repair and operate the rapidly growing repertoire of complex machines made and operated by the American factory system. From the tiny wrenches used to switch a brush blade with a grass blade on a factory-made scythe to the wide variety of strange looking wrenches used to operate wagons and 19th century farm equipment, the wrench replaced the edge tool as an icon of the rise of the machine age (read "Iron Age") in the context of the gradual end of the Wooden Age (Figure 608 to Figure 632).

One may postulate a second stage of the factory system of mass production, where increasingly sophisticated "steel" machinery replaced cast iron machinery. This next stage in our cascading series of Industrial Revolutions, which is noted in this text as the Age of Steel (Figure 743), is characterized by the introduction of tempered alloy steels and the continued perfection of grey cast iron and ductile steel for tool production. The increase in production of wrenches in a variety of forms coincided with the florescence of now famous manufacturers of machinist tools such as the L. S. Starrett Co. and the Brown & Sharpe Co. In reality these historical events were more like an exuberant square dance, with the wrench-maker partnered with the tool and die maker, who made the hand tools that built the machinery operated and adjusted by the products of the wrench-maker.

Many other tradespersons and artisans made their appearance in the last years of the 19th century to repair and maintain complex tools of all kinds. The ubiquitous plumber and heating system repairman, now among the most essential enablers of our comfortable life styles (the bad news: as the unemployed middle class becomes a dependent underclass, plumbing and heating repairs become a luxury for the elite) has an increasingly complex toolkit that now has had almost a century and a half to evolve. Many basic tools have not changed during this time frame. The electrician often uses tools which can be traced back to early American industries. His hand tools, as instruments of manual operation, remain important supplements to his myriad of meters and modern electronic equipment.

In the 20th and early 21st centuries, given the diminished role of the machinist and many other skilled workers in manufacturing industries, among the most renown of all tool wielders is the mechanic with his Snap-On, Proto, and Bonny wrench collections, used to maintain that most popular of all complex machines, the car that drives Dasein. The more sophisticated and possibly more significant toolmaker, at least until the IT revolution, the machinist, has labored in a more cerebral obscurity with his (sorry, no women allowed) Gerstner tool box, lunch pail, and daily work routine. Both are progenitors of the innovative engineers of the age of information technology. Did Steven Jobs like to repair his own automobile in his younger days?

Figure 608. Wedge adjusted wrench. Cast or forged steel. DTM. 101900T6.

Figure 609. Wagon wrench. Cast steel, rosewood. Diamond Wrench Co., Portland, Maine. CLT. 101212LTC2.

Figure 610. Alligator wrench and pipe threader. Cast steel. DTM. 6703T3.

Figure 611. Adjustable nut wrench. Drop-forged steel with cast lead handle. Goodell-Pratt Co. of Greenfield, Massachusetts. DTM. 62202T6.

Figure 612. Wedge adjusted nut wrench. Drop-forged steel. Roger Printz & Co. "Fitzall." CLT. 22512LTC2.

Figure 613. Adjustable nut wrench. Drop-forged tempered alloy steel. W.J. Ladd, New York. DTM. 31212T6.

Figure 614. Ripley's patent adjustable wrench. Drop-forged steel. E. Ripley. DTM. 31501T2.

Figure 615. Combination alligator wrench and screw threader. Cast steel. Bonney Vixen. DTM. 101312T18.

Figure 616. Locking combination box wrench. Cast iron. DTM. 6703T35.

Figure 617. Handle slide adjusting wrench. Malleable iron. H. & E. Wrench Co. (G.E. Hemphill and E.J. Evans). CLT. 32412LTC1.

Figure 618. Bicycle nut wrench. Drop-forged iron and steel. Billings & Spencer. DTM. 072512T1.

Figure 619. Combination wrench and cutter. Drop-forged steel, nickel plating. CLT. 52712LTC2.

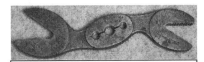

Figure 620. Combination alligator wrench and threader. Cast steel. Hawkeye Wrench Co. of Marshalltown, Iowa. CLT. 7612LTC4.

Figure 621. Jointed socket wrench with extension. Drop-forged steel. Abram W. Wheaton. CLT. 7712LTC1.

Figure 622. Offset combination tractor wrench. Forged steel. Barcalo. CLT. 7612LTC8.

Figure 623. Adjustable tractor nut wrench. Drop-forged grey cast iron. International Harvester Co. CLT. 7612LTC2.

Figure 624. Multi-socket box wrench. Drop-forged steel. Rast Product Corp. CLT. 7612LTC1.

Figure 625. Sliding jaw adjustable nut wrench. Drop-forged steel. Gellman Wrench Corporation. 7512LTC3.

Figure 626. Bryant patent alligator wrench. Drop-forged steel. Shaw Propeller Company, Boston, Massachusetts. CLT. 101212LTC4.

Figure 627. Slide adjusted nut wrench. Drop-forged steel, brass. DTM. 32708T52.

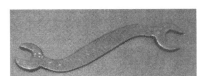

Figure 628. Curved open box wrench. Natural steel. DTM. 8912T3.

Figure 629. Offset open box wrench. Cast iron. DTM. 8912T9.

Figure 630. Offset open box wrench. Cast iron. DTM. 8912T4.

Figure 631. Offset open box wrench. Cast iron. G. Edgcumbe. DTM. 8912T2.

Figure 632. Offset crescent wrench. Cast steel, wood (rosewood). DTM. 121412T4.

Figure 633. Adjustable wrench. Drop-forged German steel, nickel-plated finish. DTM. 52603T1.

Figure 634. Adjustable pipe wrench. Drop-forged steel, wood (rosewood). F. E. Wells & Son. DTM. 52603T23.

122

Tools Manufactured by Machinery

When Robert Mushet, working in his isolated workshop in the Forest of Dean north of the Severn River in 1868, invented his "self-hard" high speed cutting tools, he laid the foundation for the evolution of an industrial society that would rapidly cut and shape machinery, equipment, and tools of every description. It was this machinery that made the equipment of the modern age: automobile engines, telephone and telegraph components, plumbing and electrical supplies, power tools (machine-made machines), marine hardware, armaments, and consumer products of every description. These high speed alloy steel cutting tools, many varieties of which were developed after 1900, supplemented and expanded the ability of machinery to drop-forge tools and equipment with interchangeable parts. The productivity of these inventions and innovations was, in turn, supplemented and enhanced by the ever growing repertoire of varieties of iron and steel, of which alloy steel and high speed tool steel cutting tools were only one variation.

As malleable iron evolved into "low-carbon" steel and "puddled" steel evolved into "tempered alloy" steel, the capacity of industrial society and its vast array of machinery to drop-forge hand tools of every description greatly expanded. By the late 19th century, the role and significance of the carbon content – the chemistry – of iron and steel were fully understood. The secrets of the ancient tedious hand-forge-welding of steeled edge tools, a high art if there ever was one, became the sophisticated ability to create tool steels of every description with subtle variations in alloy content, including the ever essential manganese, as well as tungsten, nickel, cobalt, and other alloys. High speed machinist cutting tools were not the only use of alloys, which had long been unknown components of the products of ancient ironmongers. They now became the components of machine-made tools of every description. The perfection of stainless steel by Harry Brearley in 1913, first developed by French and then German scientists between 1904 and 1912, further expanded this repertoire. The evolution of the forge-welded hand tools of early American industries progressed to the partial, and then complete, drop-forging of an ever increasing variety of hand tools. From screw augers to Hazard Knolles' first cast iron plane, the variety and metallurgical constituents of tools expanded in conjunction with the settlement of the mid, and then, far west and the rapid growth in America's industrial capacity. The classic period of American toolmaking, the zenith of machine-made tools, could aptly be described as the age of drop-forged tempered alloy steels.

Woodworking Tools

Figure 635. Traut's patent universal combination plane. Grey cast iron, tempered alloy steel, steel blades, brass nuts, with a wooden handle. Stanley. DTM. 22211T29.

Figure 636. Millers Falls boring machine. Cast iron, drop-forged malleable iron. Millers Falls Co. DTM. 82500T3.

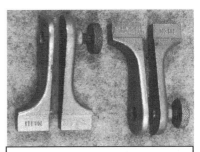

Figure 637. Stair gauge. Drop-forged steel. L.S. Starrett. CLT. 42912LTC6.

Figure 638. Stanley no. 140 Block plane. Grey cast iron, steel blade, rosewood knobs. CLT. 42812LTC4.

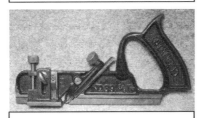

Figure 639. Stanley no. 39 3/8 dado plane. Grey cast iron, japanned finish, steel blade. CLT. 7712LTC4.

Figure 640. Stanley no. 11 belt-makers' plane. Grey cast iron, wood (beech), cast steel blade, japanned finish. CLT. 6312LTC2.

Figure 641. Stanley no. 66 beading plane. Grey cast iron with nickel-plating. DTM. 31808SLP11.

Figure 642. Archimedes drill. Steel. CLT. 5512LTC4.

Figure 643. Bailey no. F block plane. Grey cast iron, cast steel blade. Bailey Tool Co., Woonsocket, Rhode Island. DTM. 100400T1.

Figure 644. Stanley no. 12 veneer scraper. Grey cast iron, steel blade, brass fittings, wooden handle. DTM. 22211T27.

Figure 645. Stanley no. 112 carpenters' scraper plane. Grey cast iron, steel blade, brass nuts, with a wooden tote and handle. DTM. 22211T28.

Figure 646. Spiral screwdriver. Drop-forged malleable iron, brass, wooden handle. DTM. 22411T18.

Figure 647. Stanley no. 45 convex spokeshave. Drop-forged tempered alloy steel, steel blade. DTM. 22211T30.

Figure 648. Stanley no. 113 adjustable compass plane. Grey cast iron, tempered alloy steel, steel blade. DTM. 22211T26.

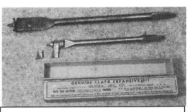

Figure 649. Clark's expansive bit. Cast steel. DTM. 21812T2.

Figure 650. Hammacher & Schlemmer hand drill. Cast iron and steel. CLT. 52612LTC3.

Figure 651. Carpenters' two speed hand drill. Drop-forged malleable iron, steel, and a wooden handle. DTM. 22211T36.

Figure 652. Dado plane. Grey cast iron, japanned finish, steel blade. Stanley. CLT. 7712LTC4.

125

Figure 653. Stanley No. 7 jointer bench plane. Grey cast iron, wood (rosewood), brass knobs, japanned finish. CLT. 7512LTC2.

Figure 654. Stanley No. 5 jack plane. Grey cast iron, wood (rosewood), brass knobs, japanned finish. CLT. 7512LTC1.

Figure 655. Tenon saw. Cast steel, brass, wood (beech). Unusual "elastic steel" mark. CLT. 8912LTC1.

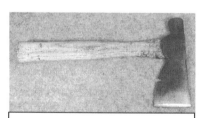

Figure 656. Stanley No. 100 tailed block plane. Cast iron, japanned finish, steel blade. CLT. 81212LTC10.

Figure 657. Stanley No. 98 side rabbet plane. Cast iron, wooden (rosewood) knob. CLT. 81212LTC9.

Figure 658. Millers Falls no. 502 corner bit brace. Cast iron, nickel plated finish, wood (cocobolo). CLT. 93012LTC1.

Figure 659. Stanley no. 1 lathing hatchet. Drop-forged steel, hickory handle. DTM. 31212T5.

Figure 660. Stanley No. 95 butt gauge. Cast iron, nickel finish. CLT. 3713LTC1.

Figure 661. Spokeshave. Brass body, steel blade. Lie-Nielsen. DTM. 111512T1.

Figure 662. Plumb bob. Turned brass and cast steel. CLT. 3213LTC4.

Figure 663. Stanley No. 130 double low angle block plane. Cast iron, steel, wood (rosewood), japanned finish. CLT. 2713LTC3.

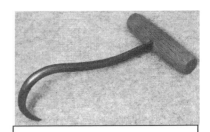

Figure 664. Cargo hook. Drop-forged steel, wood handle. C. Drew, Kingston, Massachusetts. DTM. 22612T1.

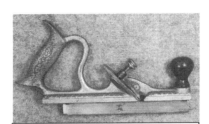

Figure 665. Stanley no. 148 match plane. Grey cast iron, nickel plated finish. CLT. 6312LTC6.

Figure 666. Stanley no. 48 tongue and groove plane. Grey cast iron body, nickel plated finish, rosewood handle, cast steel blade. CLT. 6312LTC8.

Specialized Trades of the Industrial Revolution

Railroad

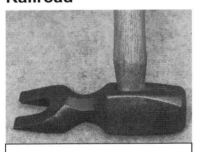

Figure 668. Railroad coal shovel. Cast steel, wood (maple). DTM. 101312T3.

Figure 667. Railroad hammer. Cast steel, wood (hickory). CLT. 10112LTC2.

Electrician

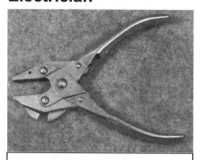

Figure 669. Wire cutting pliers. Nickel-plated drop-forged steel. Sargent & Co. CLT. 42912LTC10.

Figure 670. Button patent wire pliers. Drop-forged steel. CLT. 42912LTC11.

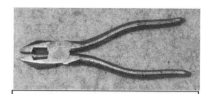

Figure 671. Linemans' pliers. Drop-forged iron and steel. The Waymouth Corp., Pawtucket, Rhode Island. DTM. 072112T8.

Figure 672. Linemans' splicing clamp. Drop-forged steel. Klein & Sons. CLT. 72712LTC6.

Figure 673. Battery pliers. Drop-forged steel. Pazzano Wrench Co., Waltham, Massachusetts. CLT. 6312LTC5.

Plumber

Figure 674. Gasoline blowtorch. Drop-forged steel, brass. CLT. 72712LTC12.

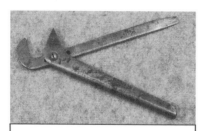

Figure 675. Pipe tongs. Cast steel. CLT. 72712LTC11.

Figure 676. Mechanics' cold chisel. Drop-forged steel. DTM. 93011T8.

Figure 677. Angle locknut pliers. Cast steel. DTM. 101312T19.

Figure 678. Adjustable pipe wrench. Drop-forged iron, brass, and wood. DTM. 32802T4.

Figure 679. Lead pipe expansion pliers. Steel. DTM. 102612T11.

Figure 680. Adjustable pipe tongs. Steel. Jarecki Mfg. Co., Erie, Pennsylvania. DTM. 102612T10.

Figure 681. Adjustable pipe wrench. Cast steel, red paint. The Ridge Tool Co., Elyria, Ohio. DTM. 121412T18.

Figure 682. Bolt cutters. Drop-forged steel. John W. Geddes and H.K. Porter of Everett, Massachusetts. CLT. 51012LTC4.

Figure 683. Offset slip joint pliers. Drop-forged tool steel. H.D. Smith & Co. of Plantsville, Connecticut. CLT. 52712LTC4.

Figure 684. Combination pipe and nut wrench. Cast steel, rosewood, black paint. Bemis & Call. CLT. 2713LTC5.

Other Factory-Made Tools

Figure 685. Lace cutter. Drop-forged iron and steel, japanned finish. DTM. 111001T34.

Figure 686. Stanley No. 30 angle divider. Drop-forged steel, nickel finish. CLT. 2713LTC4.

Figure 687. Book press. Cast iron, japanned finish. CLT. 72312LTC1.

129

Figure 688. Cobblers' lasting pliers. Drop-forged steel. Whitcher. DTM. 3312T2.

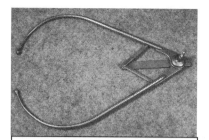

Figure 689. Medical calipers. Drop-forged steel. Willis, Philadelphia, Pennsylvania. CLT. 42912LTC5.

Figure 690. Upholsterers' shears. Cast steel, brass. CLT. 72712LTC8.

Figure 691. Banana knife. Steel, hardwood handle. CLT. 42712LTC3.

Figure 692. Pliers for hog ringing. Drop-forged German steel. Silas & John W. Sparks. DTM. 102512T12.

Figure 693. Shingle rip. Drop-forged iron and steel. C. Drew & Co. DTM. 51212T1.

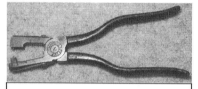

Figure 694. Leather eyelet tool. Drop-forged steel. Eyelet Tool Co., Boston, Massachusetts. DTM. 22612T2.

Figure 695. Putty knife and scraper. Drop-forged malleable iron or steel blade, rosewood handle. John Russell & Co. Green River Works, Greenfield, Massachusetts. DTM. 3312T16.

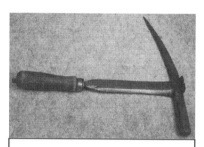

Figure 696. Slaters' hammer. DTM. 63012T2.

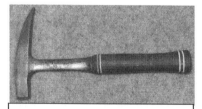

Figure 697. Rock hammer. Drop-forged steel, compressed leather discs. Estwing. DTM. 101312T29.

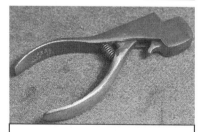

Figure 698. Conductors' railroad punch. Drop-forged steel, nickel-plated finish. McBee. CLT. 72712LTC7.

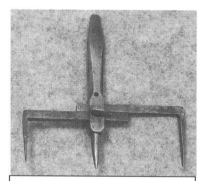

Figure 699. Leather washer cutter. Drop-forged iron and steel. C.S. Osborne & Co. CLT. 42912LTC16.

130

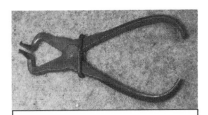

Figure 700. Tire pliers. Drop-forged steel. CLT. 42912LTC12.

Figure 701. Valve lifter. Drop-forged malleable iron. CLT. 42912LTC14.

Figure 702. Pliers. Drop-forged steel. DTM. 6708T8.

Figure 703. Rock chisel. Forged iron and steel. P. Delvin. DTM. 121112T7.

Figure 704. Socket chisel. Steel. Eric Berg. DTM. 102512T15.

Figure 705. Hand vise. Forged steel, brass ferrule, wood (cocobolo). DTM. 42912LTC2.

Figure 706. Taxidermy knife. Cast steel, stag horn, ivory, silver trappings. Warren Anatomical Museum, Boston, Massachusetts. CLT. 62112LTC2.

Figure 707. Screw adjusted locking nut wrench. Cast Steel. DTM. 7602T6.

Figure 708. Nail puller. Cast iron. CLT. 5512LTC1.

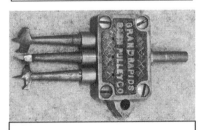

Figure 709. Three bit gang drill. Cast iron, cast steel. Grand Rapids Sash Pulley Co. CLT. 101212LTC1.

Figure 710. Right: Riveting machine. Malleable cast iron, japanned finish. F.H. Smith Mfg. Co. CLT. 72712LTC9.

Tools as Sculpture Objects

If not used or needed for utilitarian purposes, tools can be boring. Laid aside and forgotten, they become the accidental durable remnants in abandoned workshops, cobwebbed cellars, and forlorn tool chests destined for the dump. If not explored or reopened and inspected by some tool picker, they are often the subject of a call for the services of Mr. Junkit and his ubiquitous trucks. Some found tools are collectible. Their rediscovery elates hoarders who have no interest in function or craftsmanship. As wannabe klepto-plutocrats, the potential financial value of a flea market collectible has brought a smile to the face of many a scouring hoarder. Not often seen at the flea market fair are the more knowledgeable tool enthusiasts who celebrate the marriage of tools, art, and history. Artisans, technicians, teachers, collectors, etc., are all alert to the crafts-making and historical significance of the legacy of finely made American hand tools.

Found tools as accidental durable remnants also have another dimension: as sculptural objects sometimes eliciting that indescribable epiphany characteristic of the response that accompanies an encounter with a work of art. The medium doesn't matter though it might be the message. An epiphany, and oh yes, who created that work of art? This philosophical component of simple machines and Tools Teach paradigms is explored in the museum's publication *The Phenomenology of Tools* (Brack 2010). Some of our favorite tools are illustrated in this text, the iconography of which, as with the pleasures of their use, cannot be fully communicated on the printed page.

Some of these tools are still occasionally used by craftsperson or artists; many are collectibles we have encountered. Among the most unique forms, many characteristic of specialized trades, or with unusual and sometimes fleeting functions, include the following.

Figure 711. Card stretcher. Steel, nickel finish. DTM. 42912T13.

Figure 712. Inclinometer. Drop-forged iron and steel, brass. Davis Level & Tool Co. DTM. 120907T2.

Figure 713. Birmingham plane. Grey cast iron, steel. DTM. TJE1002.

Figure 714. Standard Rule Co. smooth plane. Grey cast iron, cast steel, wood. DTM. TJE4001.

Figure 715. Bronze goat face hammer head. Bronze. DTM. 22211T6.

Figure 716. Stanley no. 43 adjustable plow plane. Grey cast iron, brass, wood (rosewood). DTM. 122912LTC1.

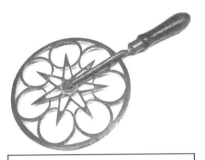

Figure 717. Fancy star pattern wheelwrights' traveler. CRF. Skowhegan Tool Museum, 1998.

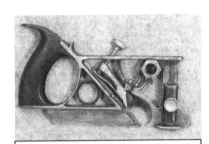

Figure 718. Phillips patent plough plane. Cast iron, cast steel, japanned finish. CLT. 10112LTC1.

Figure 719. Organ tuning cone. Lathe turned steel. DTM. 12413T1.

Figure 720. Crooked knives. CRF. Private sales.

Figure 721. Railroad spike puller. CRF. Hulls Cove Tool Barn, 2012.

Figure 722. Boston metallic rabbet plane. CRF. Private sale, 2007.

**Figure 723. Birmingham 6"
metallic rabbet plane. CRF.
Nashua auction parking lot
sale, 1995.**

**Figure 724. Bouffard patent
scraper. CRF. Ebay, 2010.**

**Figure 725. Folding safety ax.
CRF. Fairfield auction, 2005.**

**Figure 726. Combination
hammer and wrench. C.J.
Maggard. CRF. 1902 patent
#711,408. eBay, 2012.**

**Figure 727. Morris 1870 patent
plane. CRF. Brown's Auction,
2001.**

**Figure 728. Leonard Bailey
split frame plane. CRF. Private
sale, 1999.**

**Figure 729. Stanley No. 50
combination plane. CRF.
Private sale, 2005.**

**Figure 730. Stanley No. 41
combination plane. CRF. eBay,
1999. Type 1 with hook.**

**Figure 731. Steers patent
Brattleboro Tool Co. block
plane. CRF. Private sale, 2009.**

Figure 732. Birmingham rabbet plane. CRF. Nashua, NH tool flea market, 2001.

Figure 733. Lee's patent stop chamfer plane. CRF. From a tool picker who bought it for $250 at Brimfield, MA antique sale, 2005.

Figure 734. Lowentraut hammer and wrench combination tool. CRF.

Figure 735. Fox scraper. CRF.

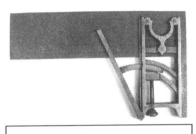

Figure 736. Square. J.M. Clouse. CRF.

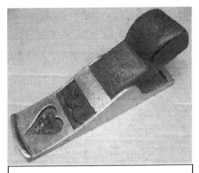

Figure 737. Heart block plane. CRF.

Figure 738. Inventor's combination tool with original box. CRF. Martin Donnelly Auctions, 2001.

Figure 739. Level. CRF. Marked Pat. Apl'd For. Ebay, 2010.

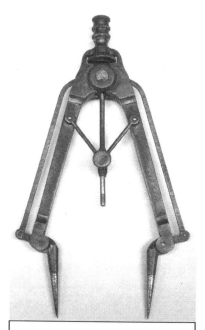

Figure 741. Double bevel. St. Johnsbury. CRF. Houston Brooks Auctions, 1998.

Figure 740. Dividers. Bronze. C. Albert. CRF. Patent #341,714. Ebay, 2001.

Figure 742. Ornamental hammer. Charles Henning. CRF. 1901 patent #D35100. Houston Brooks Auctions, 2005.

Cascading Industrial Revolutions

Most early American industries, such as the tinsmith, cooper, sail-maker, farrier, and wheelwright, have disappeared in the cascade of multiple Industrial Revolutions. Yet many of their tools linger as testaments to forgotten trades and are often adopted for various contemporary uses by artisans and craftspersons. Obscure museum collections and historical reenactments provide a reminder of their historical legacy.

Both the wrench wielder and machinist as toolmaker remain as the co-enablers of the age of petrochemical-electrical man, progenitors of our now rapidly expanding, if grossly inequitable, global consumer society. Unfortunately for machinists, although less so for mechanics, a sixth Industrial Revolution (Figure 743) has occurred, beginning in the 1980s with the development of the microchip and the birth of the Age of Information Technology. Computer-driven industrial manufacturing systems, including modern CNC technology (computer numerical calculations) has eliminated the need for millions of factory workers, whose hands-on expertise and manual labor has been replaced by Computer Assisted Design (CAD) programs to manufacture everything from automobiles, machinery, and computers, to consumer products of every description. Only a small number of highly skilled technicians are needed to maintain and operate the infrastructure of CNC technologies. The age of petrochemical-electrical man has thus evolved into the age of information technology (IT). Large numbers of workers, many of whom are underemployed or unemployed, have the frustrating function of being consumers of the products of CNC technologies while suffering the consequences of the adoption of IT: lagging incomes as the result of the lack of specialized skills. The 2012 presidential debates did not address the impact of the information technology revolution, which despite the prosperity of the techno-elite, results in unemployment and underemployment in all developed and developing countries. The earlier rise of an atomic age of nuclear powered steam turbines and atomic weapons (1945-1990) has now merged with the Age of Information Technology to further complicate the predicament of modern society, which now faces the limiting factor of finite round world natural resources, especially potable water, and the growing economic impact of overpopulation and cataclysmic climate change.

The increasing challenges of formulating sustainable economies and lifestyles mandate that we not forget the tools and trades that built a nation prior to the looming cataclysm of biocatastrophe. Participants in future sustainable societies will inevitably discover that knowledge of the hand tools of the past as well as contemporary hand tools, and a concurrent capability of utilizing these hand tools, is an essential ingredient of independent and convivial, rather than codependent and subservient, lifestyles of the future. Understanding the historical context of the invention and use of hand tools as prime mover of human social development before and during the ongoing cascade of Industrial Revolutions will be a key element of the sustainability of future economies.

The study of the history of technology is also relevant to the contemporary educational paradigm that knowledge of math and science will be an important component in maintaining the future viability of modern information technology driven consumer society. Knowledge of the history of science and technology, whether chemistry, physics, or industrial history, is an essential component of expertise in any aspect of modern information and communications technology. Inevitably, the history of technology and the evolution of industrial society and its wide variety of tool forms must be a component of public education as we struggle to remain a world leader in technological productivity and innovation. Failure to include a study of the roots of early American industries and their evolution into the machine age and the continuing role of tools in the age of information technology leaves out some of the most interesting chapters in the history of science. This would be a major omission for any viable educational program.

We are now entering the age of biocatastrophe characterized by social unrest and political resentment. In this era of declining public and natural resources, increasing humanitarian needs, and ballooning sovereign and personal debt, we can look back and admire the creative achievements of America's factory system of tool production, an industrial age that for the short span of a century (1850 – 1950) made the manufacturing of hand tools an art form. As we turn back in time to the era of early American industries, we can explore the roots of our industrial florescence and the creative achievements of the shipsmiths, edge toolmakers, ironmongers, and woodworkers who "grew" America in the preceding two centuries. The history of technology and the tools that characterize the cascading Industrial Revolutions of the past are a fascinating subject of study for any interested student. More important is the legacy, and often the physical presence and thus availability, of these tools and their relevance to the challenges of establishing and maintaining sustainable economies in our evolving age of biocatastrophe. As we confront the reality of living in a global consumer society characterized by growing income disparities, a predatory shadow banking network, and an increasingly powerful and entrenched klepto-plutocracy, the iconography – the images – of American hand tools narrate important stories about the history of a finite round world biosphere we all inhabit.

Figure 743. The Cascading Industrial Revolutions of Pyrotechnic Industrial Society

Historical Era	Date	Event
Pre-Industrial, prime mover: animal power	1900 BC	First production of high quality steel edge tools by the Chalybeans from the iron sands of the south shore of the Black Sea
	1200 BC	Steel is probably being produced by the bloomery process
	800 BC	Carburizing and quenching are being practiced in the Near East
	800 BC	Celtic metallurgists begin making natural steel in central and eastern Europe
	650 BC	Widespread trading throughout Europe of iron currency bars, often containing a significant percentage of raw steel
	400 BC	Tempered tools and evidence for the 'steeling' of iron from the Near East
	300 BC	The earliest documented use of crucibles for steel production was the smelting of Wootz steel in Muslim communities (Sherby 1995)
	200 BC	Celtic metallurgists begin supplying the Roman Republic with swords made from manganese-laced iron ores mined in Austria (Ancient Noricum)
	55 BC	Julius Caesar invades Britain
	50 BC	Ancient Noricum is the main center of Roman Empire ironworks. Important iron producing centers are also located in the Black Mountains of France and southern Spain
	43-410	Romans control Britain
	125	Steel is made in China by 'co-fusion'

Historical Era	Date	Event
	700	High quality pattern-welded swords being produced in the upper Rhine River watershed forges by Merovingian swordsmiths from currency bars smelted in Austria and transported down the Iron Road to the Danube River
	1000	First documented forge used by the Vikings at L'Anse aux Meadows (Newfoundland)
	1021	Roots of the scientific method developed via Alhazen's innovations in mechanical physics and optics
	1040	Bi Sheng invents interchangeable moveable type press
	1250	First blast furnace built in Lapphyttan, Sweden
	1300	Cannon and German steel production from blast furnaces begins
	1350	Widespread appearance of blast furnaces in central and northern Europe
	1400	Gun production becoming widespread
	1439	Johannes Gutenberg develops the first printing press
	±1465	First appearance of blast furnaces in the Forest of Dean (England)
	1500	Systemic exploration and settlement of the New World
	1509	Natural steel made in the Weald (Sussex, England) by fining cast iron
	1560	First floating docks in Venice, Venetian Republic
Proto-Industrial Revolution, prime mover: wooden sailing ships	**1601**	First record of the cementation process, in Nuremberg
	1605	First commercial production of a newspaper in Europe
	1607	First shipsmith forge in the American colonies used at Fort St. George, Maine
	1613	Cementation process is patented in England to produce blister steel
	1620	Settlement of Plimouth Plantation
	1625	First Maine shipsmith, James Phipps, working at Pemaquid
	1629-1642	The great migration of Puritans from England brings hundreds of trained shipwrights, shipsmiths, and ironworkers to New England
	1646	First colonial blast furnaces and integrated ironworks are established at Quincy and Saugus, Massachusetts
	1652	James Leonard establishes the first of a series of southeastern Massachusetts colonial era bog iron forges on Two Mile River at Taunton, Massachusetts
	1661	Invention of chemistry
	1675-1676	King Philip's War in southern Massachusetts and Rhode Island
	1676	The great diaspora (scattering) of Maine residents living east of Wells follows the King Philip's War
	1686	Widespread use of the cementation process in England to produce blister steel begins

Historical Era	Date	Event
	1689-1697	The war of the League of Augsburg
	1702-1714	The war of Spanish Succession
	1703	Joseph Moxon ([1703] 1989) publishes *Mechanick Exercises or the Doctrine of Handy-Works*
	1709	Abraham Darby discovers how to use coke instead of coal to fuel a blast furnace
	1713	First appearance of clandestine steel cementation furnaces in the American colonies produces blister steel for toolmaking
	1720	First of the Carver, Massachusetts blast furnaces established at Popes Point
	1720	William Bertram invents manufacture of 'shear steel' on Tyneside
	1722	René de Réaumur (1722) provides the first detailed European account of malleableizing cast iron
	±1742	Benjamin Huntsman adapts the ancient process of crucible steel-production for his watch spring business in Sheffield, England
	1745	The first electrical capacitor, the Leyden jar, invented in the Netherlands
First Industrial Revolution, prime mover: steam engines	**1750**	Widespread use of coal, then coke, reinvention of cast steel
	1754-1763	The Seven Years War in Europe results in the last of the French and Indian wars in eastern North America
	1758	John Wilkinson begins the production of engine cylinders made with the use of his recently invented boring machine
	1759	The defeat of the French at Quebec by the English signals the end of the struggle for control of eastern North America
	1762	Jean-Baptiste Vaquette de Gribeauval begins designing artillery and guns with interchangeable parts
	1763	The Treaty of Paris opens up eastern Maine for settlement by English colonists
	1763-1769	James Watt designs and patents an improved version of the Newcomb atmospheric engine, i.e. the steam engine
	1774	John Wilkinson begins the mass production of engine cylinders used in Watt's steam engine pressure vessels
	1775	Matthew Boulton and James Watt begin mass production of steam engines
	1775	The American Revolution begins
	±1783	The approximate date when Josiah Underhill began making edge tools in Chester, NH. The Underhill clan continued making edge tools in NH and MA until 1890
	1783	James Watt improves the efficiency of the steam engine with introduction of the double-acting engine

Historical Era	Date	Event
	1784	Henry Cort introduces his redesigned reverbatory puddling furnace, allowing the decarburization of cast iron to produce wrought and malleable iron without contact with sulfur containing mineral fuels
	1784	Henry Cort invents and patents grooved rolling mills for producing bar stock and iron rod from wrought and malleable iron
	1789-1807	Era of great prosperity for New England merchants due to the neutral trade
	1792	Eli Whitney presents the idea of muskets with interchangeable parts in the USA
	1793	Samuel Slater begins making textiles in Pawtucket
	1795	Reverbatory furnace, rolling mills
	1800	Alessandro Volta invents the voltaic pile, the first battery; High pressure steam engine developed by Richard Trevithick in England
	1801	Jacquard loom, first punch card program operated machine
	1802	Incandescent light bulb (Davy)
	1802-1807	Henry Maudslay invents and produces 45 different types of machines for mass production of ship's blocks for the British Navy
	1804	First appearance of a railway steam locomotive in England; invented by Richard Trevithick
	1804	Samuel Lucas of Sheffield invents the process of rendering articles of cast iron malleable
	1813	Jesse Underhill is first recorded as making edge tools in Manchester, NH
	1815-1835	The factory system of using interchangeable parts for clock and gun production begins making its appearance in the United States
	1818	Thomas Blanchard designs a pattern tracing lathe for turning irregular gunstocks
	1820	Steam-powered saw mills come into use near Bath, Maine, shipyards
	1822	Charles Babbage invents the difference engine
	1826	Thomas Blanchard develops the first steam powered car
	1828	Adoption of the hot air blast improves blast furnaces
	1831	Seth Boyden of Newark, NJ, first produces malleable cast iron commercially in the US
	1832	D. A. Barton begins making axes and edge tools in Rochester, NY
	1832-1853	Joseph Whitworth introduces innovations in precision measurement techniques and a standardized decimal screw thread measuring system

Historical Era	Date	Event
	1835	Malleableized cast iron is first produced in the United States
	1835	Steel is first made by the puddling process in Germany
	1835	The first railroad is established between Boston and Worcester, Massachusetts
	1837	The Collins Axe Company in Collinsville, Connecticut, begins the production of drop-forged axes
	1837	In England, Joseph Nasmyth introduced the steam-powered rotary blowing engine
	1837	Commercially produced DC motors introduced by Sturgeon and Davenport
	1839	William Vickers of Sheffield invents the direct conversion method of making steel without using a converting furnace
Age of Iron, prime mover: machinery	**1840**	Factory system of mass production begins
	1842	Joseph Nasmyth patents his steam hammer, facilitating the industrial production of heavy equipment, such as railroad locomotives
	1843	The first example of computer programming invented by Lovelace
	1849	Thomas Witherby begins the manufacture of chisels and drawknives in Millbury, MA
	1850	Joseph Dixon invents the graphite crucible used in a steel production
	1850	Development of the railroad becomes widespread
	1851	J. R. Brown begins to manufacture a vernier caliper
	1853	John, Charles, and Richard T. Buck form the Buck Brothers Company in Rochester, NY, after emigrating from England and working for D. A. Barton. They later move to Worcester, MA in 1856 and Millbury, MA in 1864
	1856	Gasoline is first distilled at Watertown, Massachusetts
	1856	Bessemer announces his invention of a new bulk process steel-production technique at Cheltenham, England
	1857	The panic and depression of 1857 signals the end of the great era of wooden shipbuilding in coastal New England
	1857	Sound is first recorded on the Phonoautograph
	1863	First successful work on the Siemens open-hearth process
	1865	Significant production of cast steel now ongoing at Pittsburg, Pennsylvania furnaces
	1868	R. F. Mushet invents 'Self-hard,' the first commercial alloy steel
	1868	Manufacture of the micrometer caliper began in America
	1870-1885	Era of maximum production of Downeasters in Penobscot Bay (large four-masted bulk cargo carriers)
	1870	Tempered alloy steel production becomes widespread

Historical Era	Date	Event
	1874	Tilting band saw is introduced and revolutionizes shipbuilding at Essex, MA
Age of Steel, prime mover: railroads	**1875**	Bulk processed steel via the Bessemer process now produced in large quantities
	1876	Alexander Bell patents the telephone
	1879	Sidney Gilchrist Thomas invents basic steelmaking
	1880	Thomas Edison's electric lights become widely used
	1884	Paul Nipkow patents television
	1888	Charles Brush develops electric wind turbines for grid electricity; Nicola Tesla develops the AC motor
	1890s	Wireless data transmission, microwave and radio
Age of Petrochemical-Electrical Man, prime mover: automobiles	**1900**	
	1904	Thermionic triode vacuum tube invented
	1906	The first electric-arc furnace is installed in Sheffield; AM radio makes its first appearance
	1910	First Model-T is manufactured in Detroit
	1913	Brearley invents stainless steel
	1915	World War I, mass production of weapons, unmanned aerial drone (Tesla)
	1921	Thomas Midgley invents leaded gasoline
	1925	Electric turbines
	1926	The first high-frequency induction furnace in Sheffield
	1928	WRGB, the first television station; Thomas Midgley develops CFCs
	1930	Great Depression
	1935	Diesel engine
	1940	World War II
	1944	Colossus, the first programmable digital computer
Age of Petrochemical-Electrical-Nuclear Man, prime mover: electrical grid	**1945**	Hiroshima and Nagasaki mark the end of WWII; American consumer society begins its rapid growth
	1951	The first nuclear power plant goes online
	1955	The first transistors make their appearance
	1957	First artificial satellite (Sputnik); artificial neuroprosthetic cochlear implants developed by Andre D. Djourno and Charles Eyriès
	1959	US navy adopts unmanned aerial drones
	1960	Agricultural pesticides become widely used
	1965	First appearance of the supersonic transport
	1969	Moon landing occurs; ARPANET signals the birth of the internet
	1974	The internet becomes publically available
	1981	Scanning tunneling microscope developed
	1982	TCP/IP (modern internet protocol) appears

Historical Era	Date	Event
	1985	Digital integrated circuits become widely used
	1986	Chernobyl accident
	1990	Evolution of the global consumer society
Age of Information Technology, prime movers: fiber-optics, computers	**1995**	NSFNET decommissioned, commercialization of the internet begins
	1998	Human brain-controlled interface
	2000	Nanoscale superconductors and nanotechnology developed; meta-material production begins
	2005	Beginning of the sovereign debt crisis
	2008	World financial crisis; world hits 1 billion PCs in use
Age of Biocatastrophe	**2011**	Fukushima Daiichi; world population reaches 7 billion; cataclysmic climate change and the world water crisis become topics of widespread concern

Appendices

Appendix 1: Plates from Diderot

Diderot's *Pictorial Encyclopedia of Trades and Industry* has been republished many times. The plates in this appendix are reprinted from this version:

Diderot, Denis and d'Alembert, Jean Baptiste le Rond. [1751-65] 1964-6. *Recueil de planches sur les sciences, les arts libéraux, et les arts mechaniques avec leur explication.* 6 vols. Paris: Au Cercle du Livre Precieux.

Below are a few other versions of Diderot's *Encyclopedia* that the Davistown Museum has in its library.

Diderot, D. and d'Alembert, J. 1751-1777. *L'Encyclopédie ou dictionnaire raisonné des sciences, des arts et des métiers.* 22 vols. Paris: Braissons.

Diderot, Denis. [1751-75] 1959. *A Diderot pictorial encyclopedia of trades and industry: Manufacturing and the technical arts in plates selected from "L'Encyclopédie, ou Dictionnaire Raisonné des Sciences, des Arts et des Métiers" of Denis Diderot: In two volumes.* Vol. 1 and 2. NY: Dover Publications Inc

Diderot, Denis and d'Alembert, Jean Baptiste le Rond. [1751-75] 1969. *Pictorial Encyclopedia of Science Art and Technology* [Also called: *Compact edition: Recueil de planches sur les sciences, les arts libéraux, et les arts mechaniques avec leur explication].* Vols. 18-28. NY: Readex Microprint Corporation.

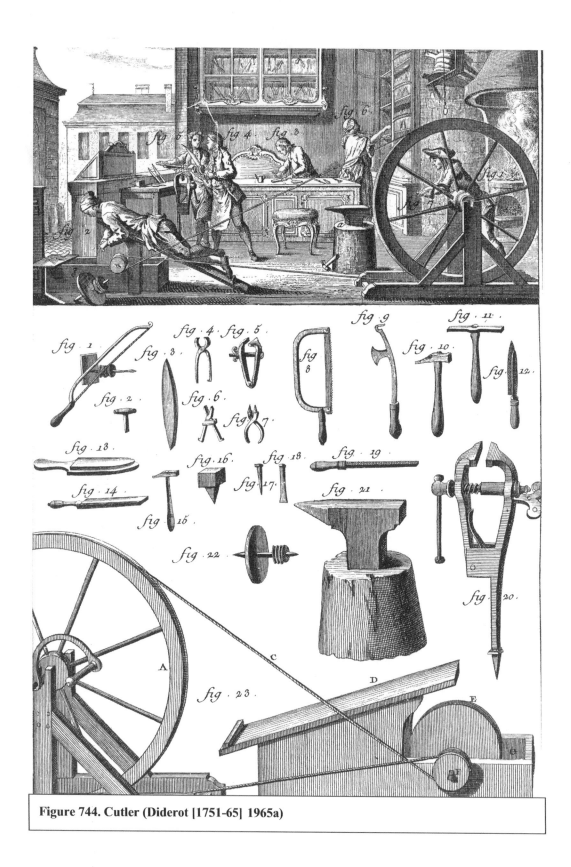

Figure 744. Cutler (Diderot [1751-65] 1965a)

146

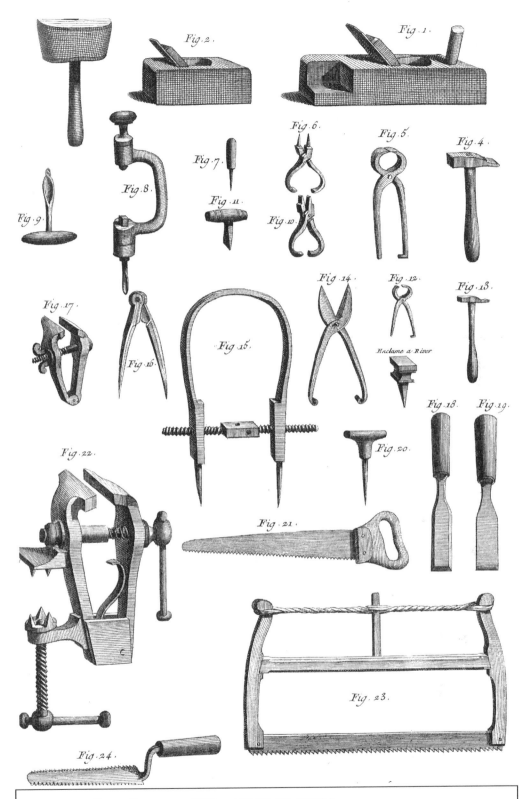

Figure 745. Crate Carpentry (Diderot [1751-65] 1965b)

147

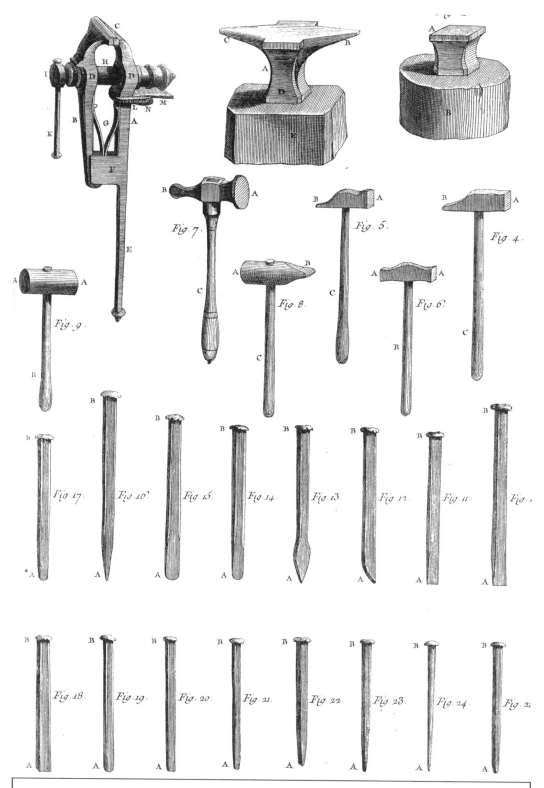

Figure 746. Swordsmith (Diderot [1751-65] 1965a)

148

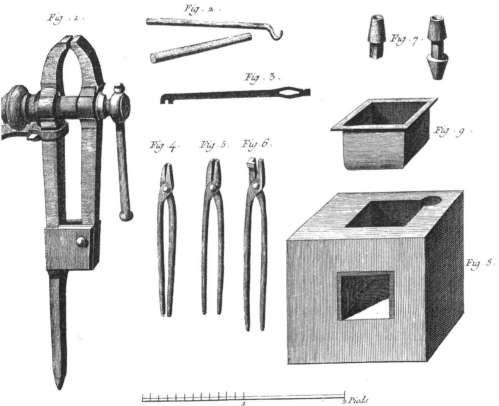

Figure 747. Blacksmith, Spurs (Diderot [1751-65] 1965a)

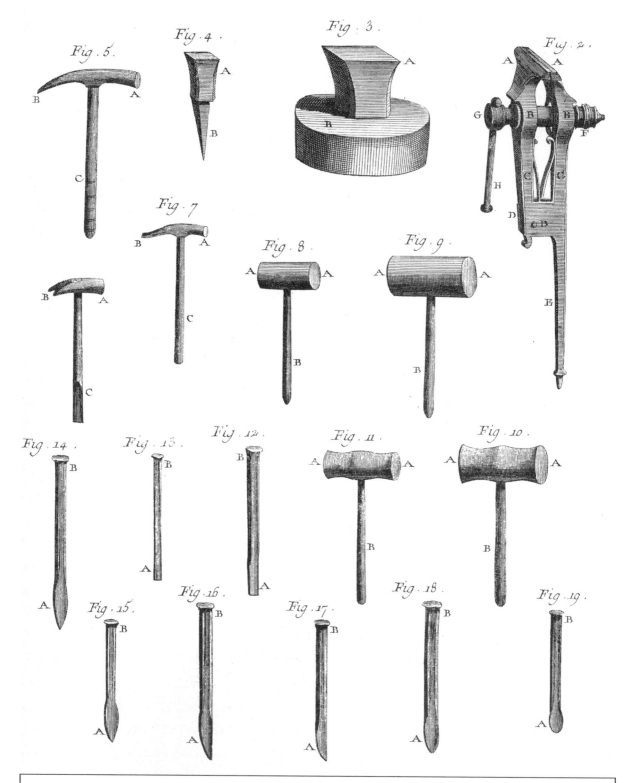

Figure 748. Carousel Saddler (Diderot [1751-65] 1965d)

150

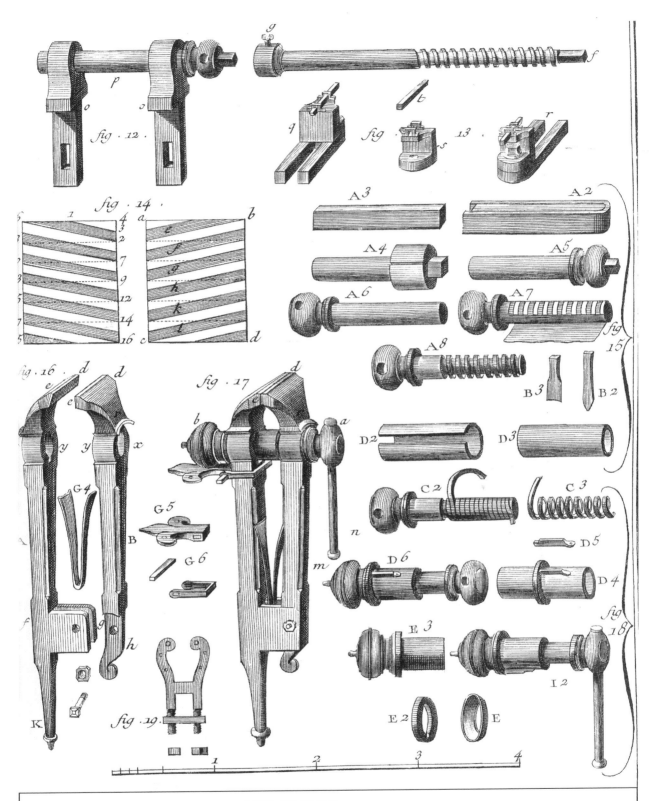

Figure 749. Metal Toolmaker (Diderot [1751-65] 1965d)

151

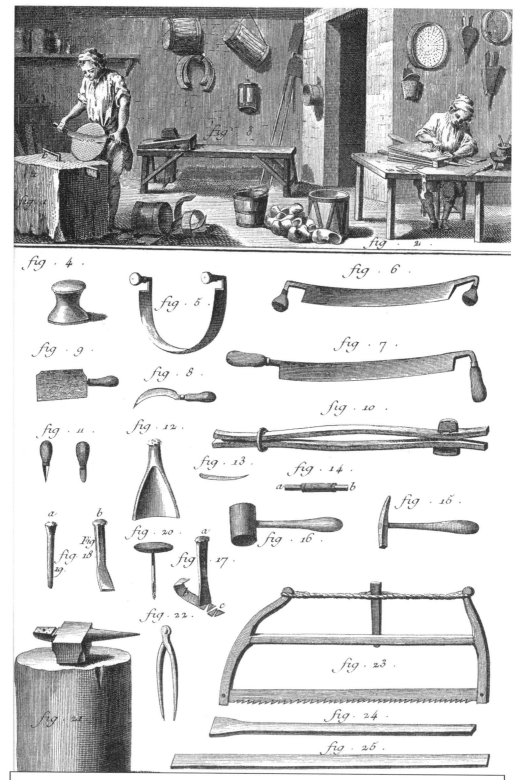

Figure 750. Cooper (Diderot [1751-65] 1964)

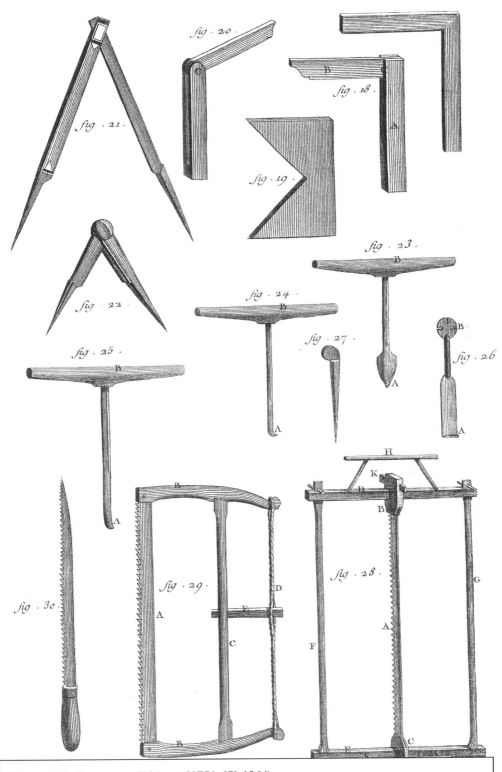

Figure 751. Carpenter (Diderot [1751-65] 1964)

153

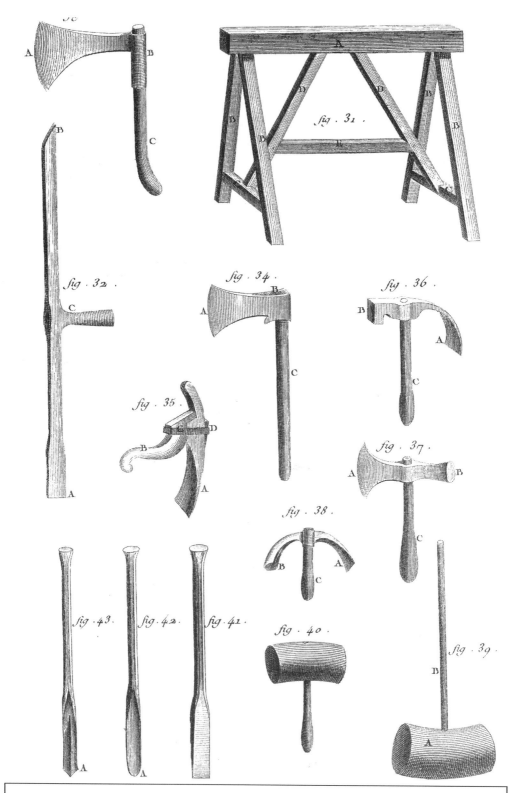

Figure 752. Carpenter (Diderot [1751-65] 1964)

154

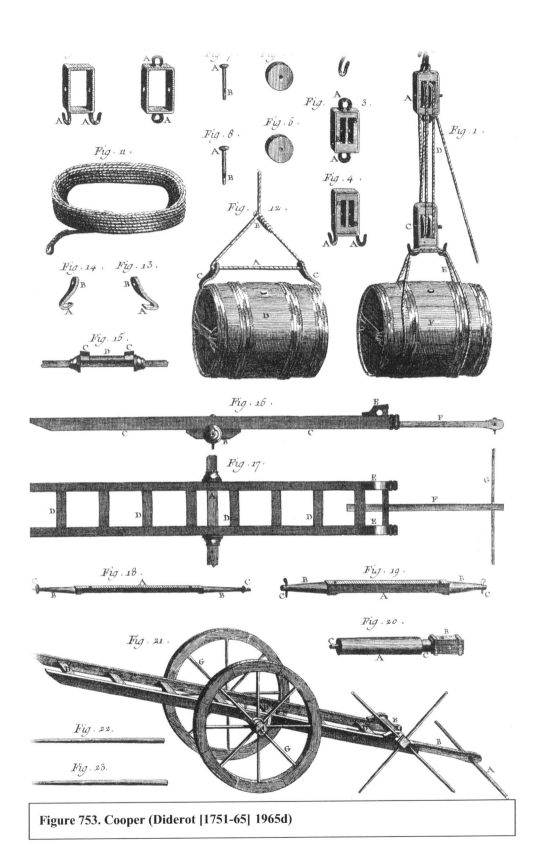

Figure 753. Cooper (Diderot [1751-65] 1965d)

155

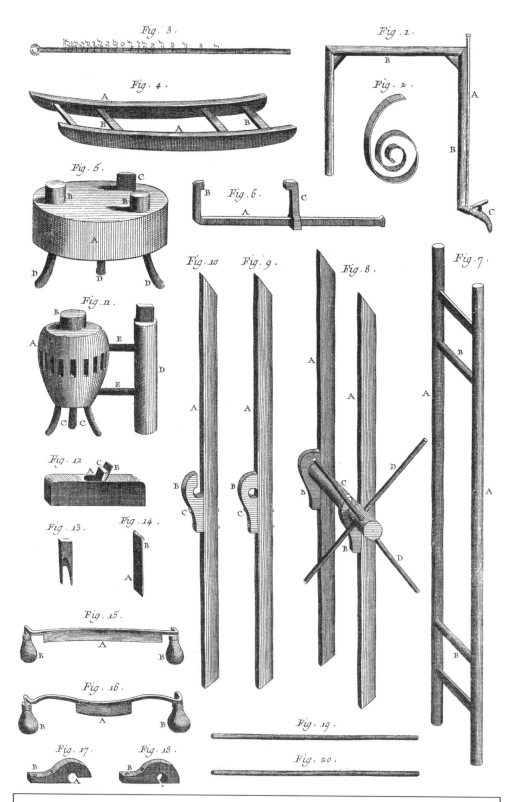

Figure 754. Cooper (Diderot [1751-65] 1965d)

156

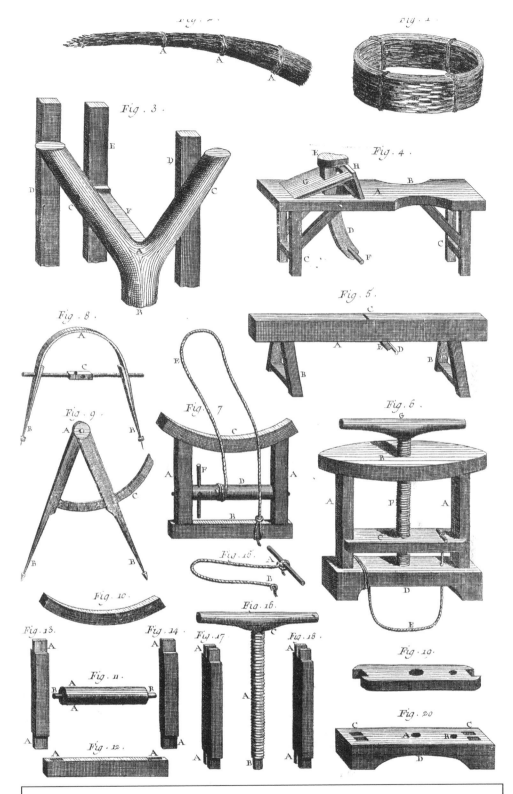

Figure 755. Cooper (Diderot [1751-65] 1965d)

157

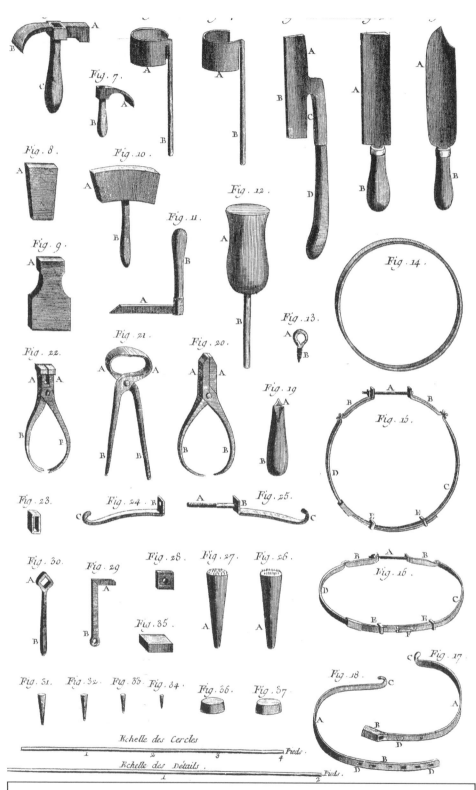

Figure 756. Cooper (Diderot [1751-65] 1965d)

158

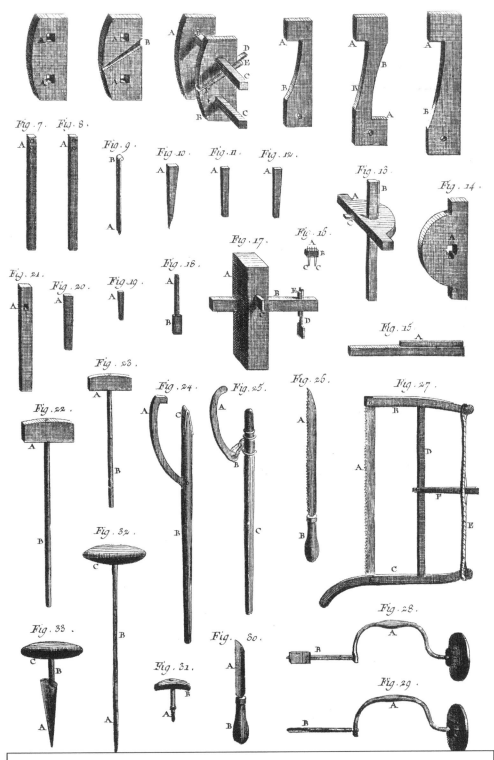

Figure 757. Cooper (Diderot [1751-65] 1965d)

159

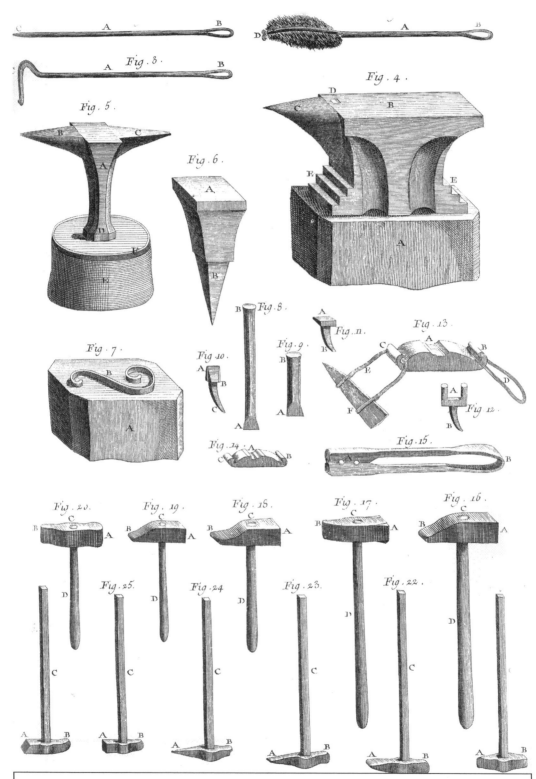

Figure 758. Locksmith (Diderot [1751-65] 1965d)

160

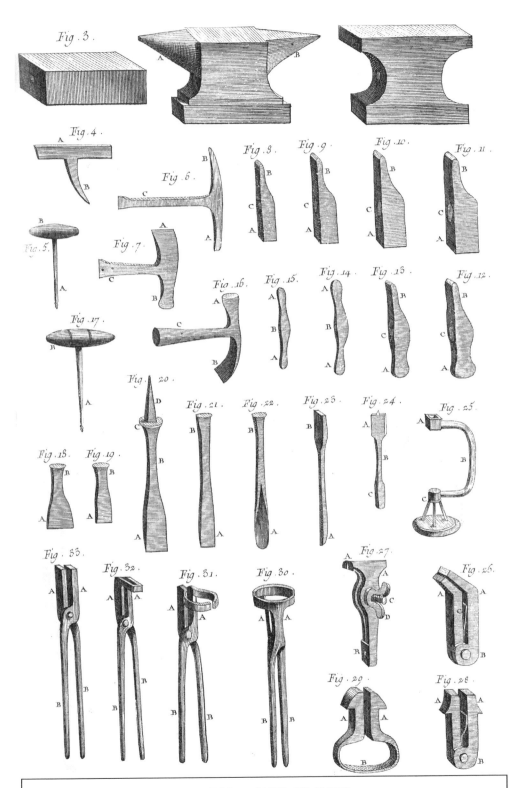

Figure 759. Toolmaker, Augers (Diderot [1751-65] 1965d)

161

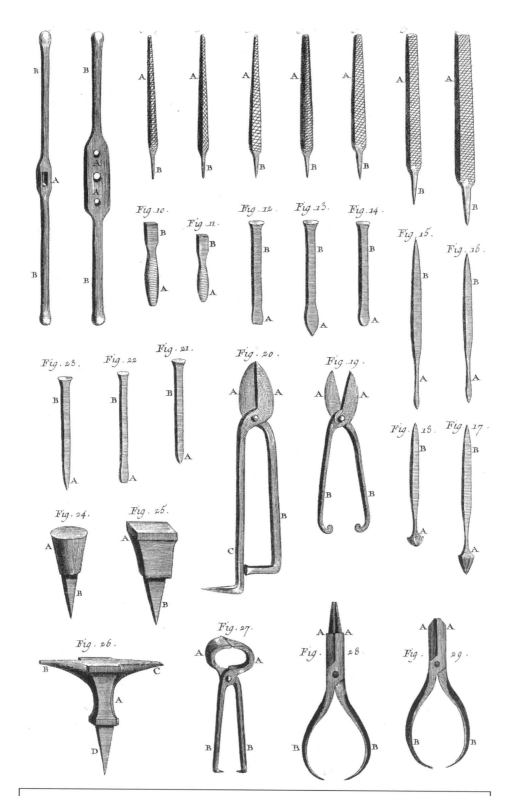

Figure 760. Toolmaker, Augers (Diderot [1751-65] 1965d)

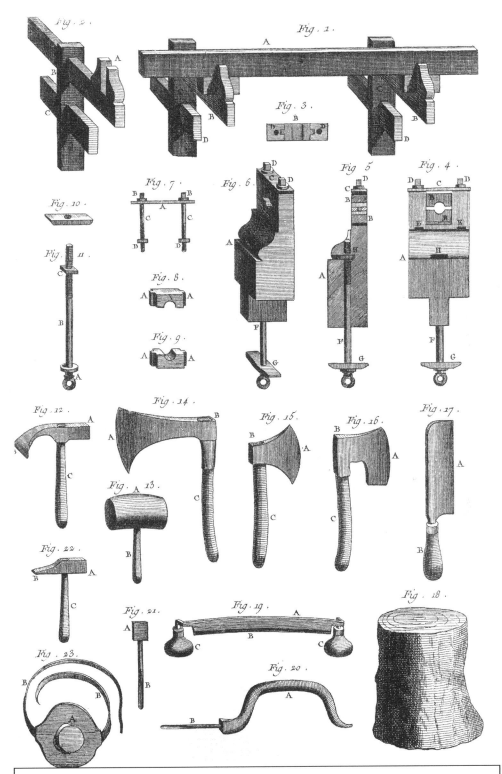

Figure 761. Turning, Wood and Iron (Diderot [1751-65] 1965d)

163

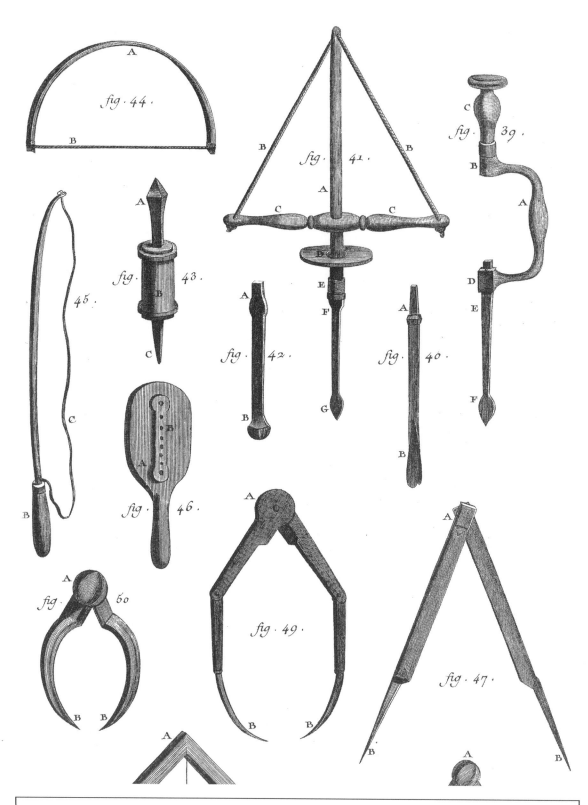

Figure 762. Masonry, Marble (Diderot [1751-65] 1965b)

164

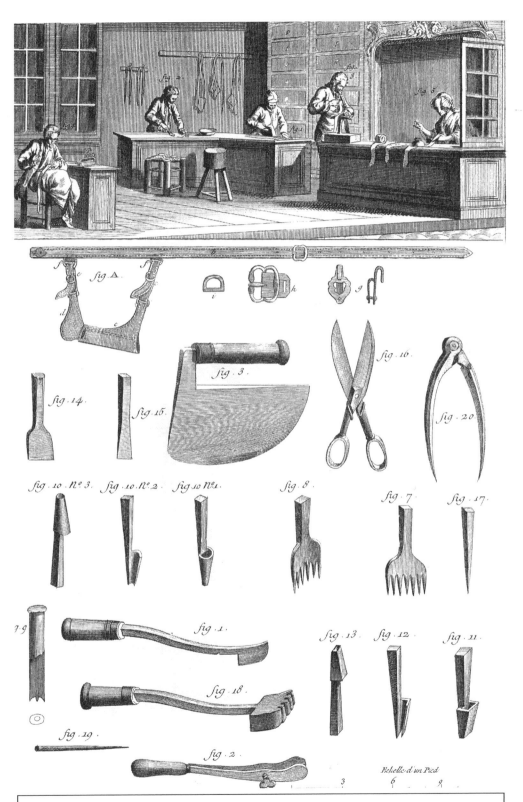

Figure 763. Leather Belt Maker (Diderot [1751-65] 1964)

165

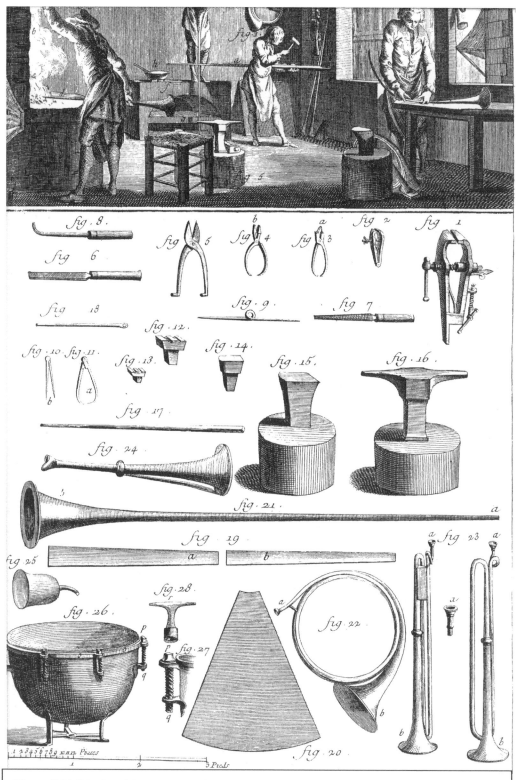

Figure 764. Blacksmith, Brass Musical Instruments (Diderot [1751-65] 1965a)

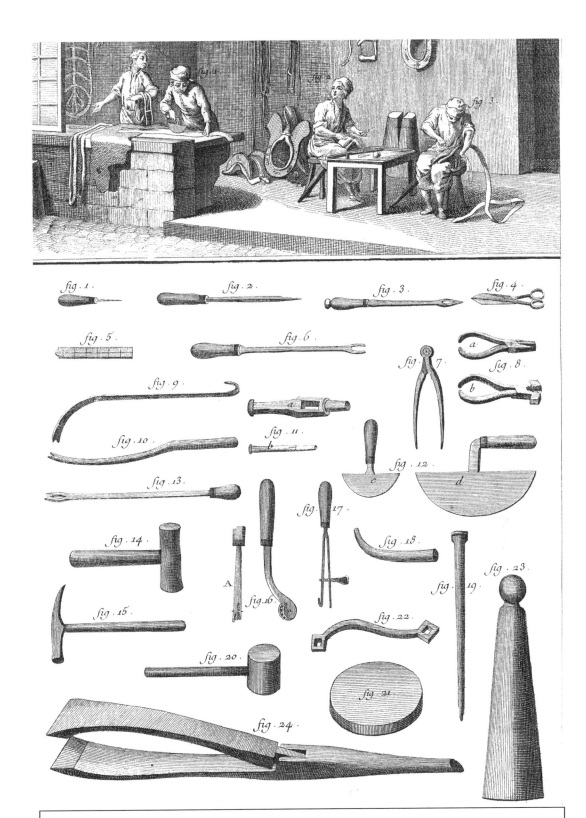

Figure 765. Saddler (Diderot [1751-65] 1964)

167

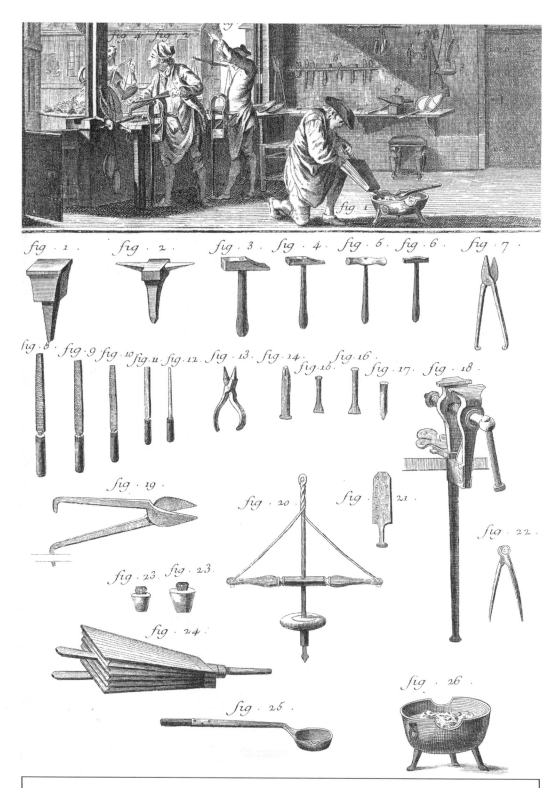

Figure 766. Scale Maker (Diderot [1751-65] 1964)

168

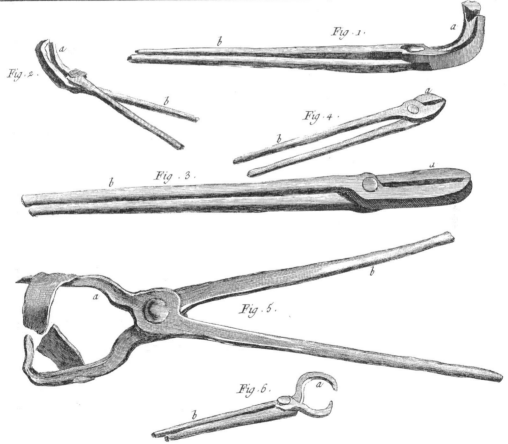

Figure 767. Wheelwright (Diderot [1751-65] 1965c)

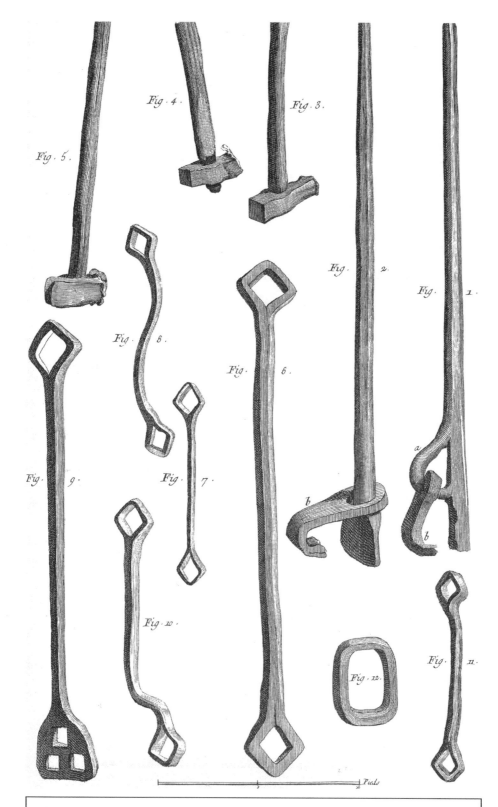

Figure 768. Wheelwright (Diderot [1751-65] 1965c)

170

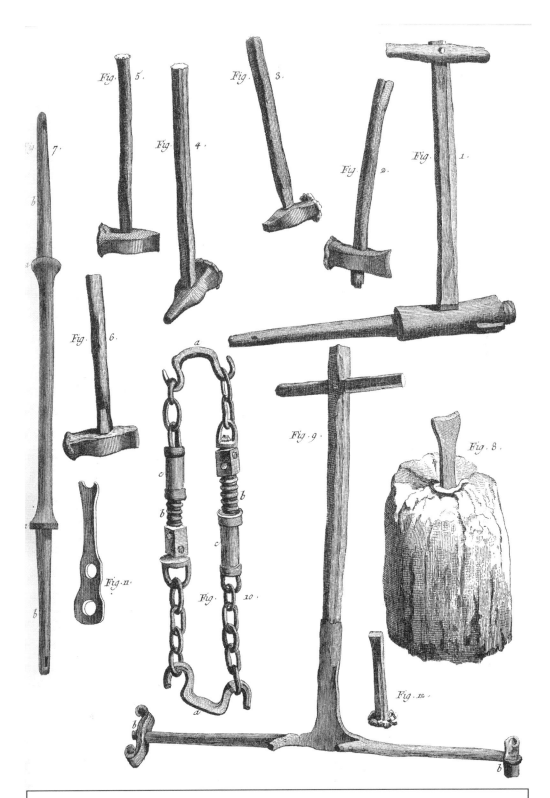

Figure 769. Wheelwright (Diderot [1751-65] 1965c)

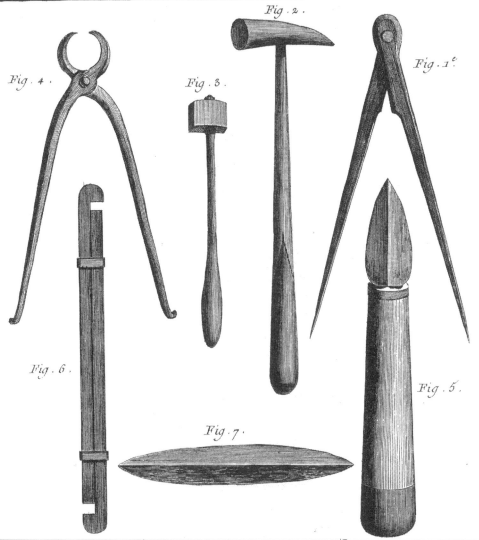

Figure 770. Glazier (Diderot [1751-65] 1965d)

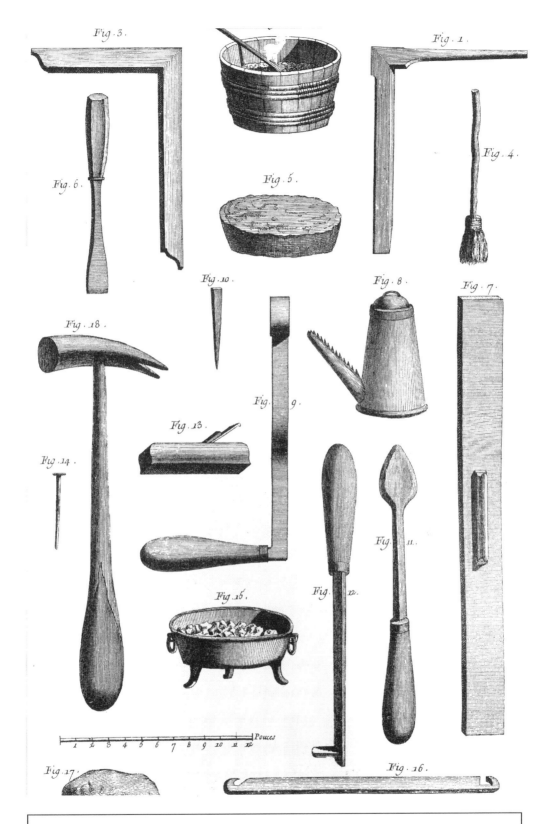

Figure 771. Glazier (Diderot [1751-65] 1965d)

173

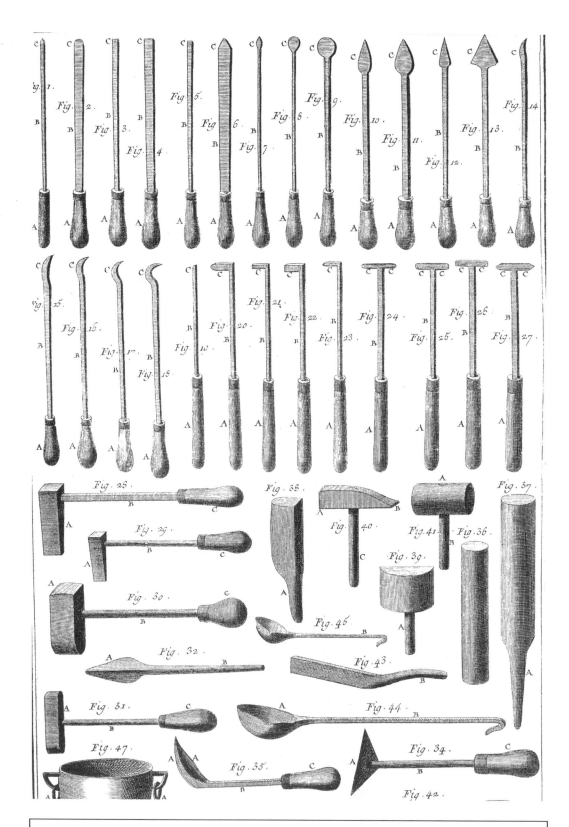

Figure 772. Tin Potter (Diderot [1751-65] 1965c)

174

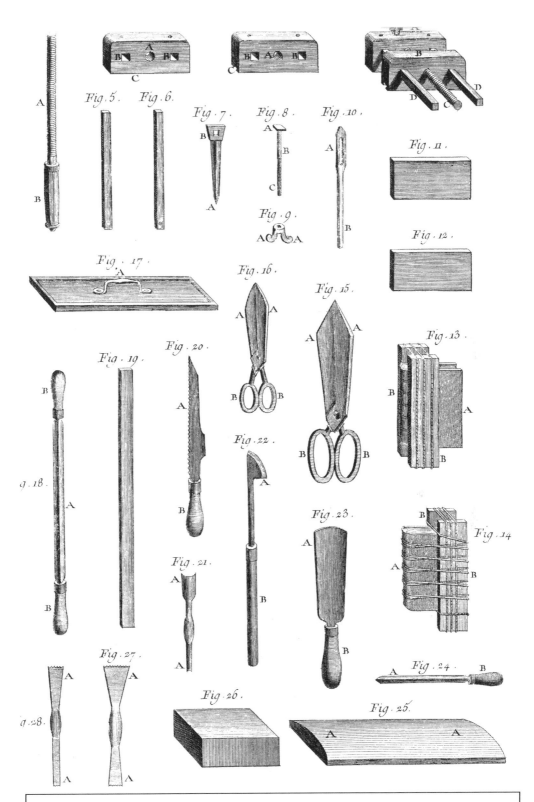

Figure 773. Book Binding (Diderot [1751-65] 1965c)

175

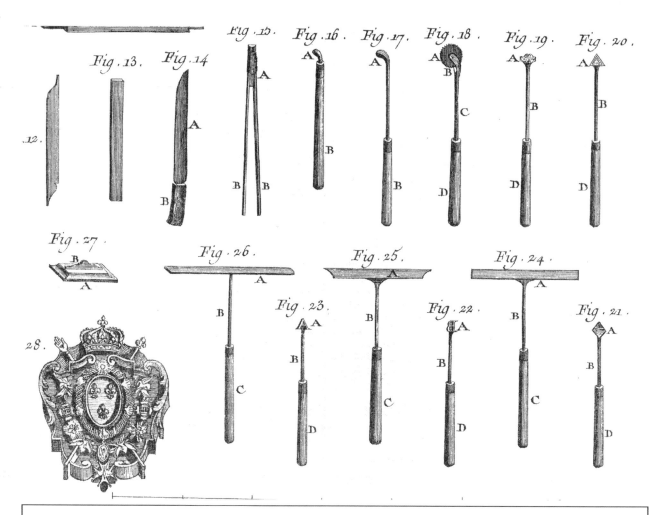

Figure 774. Book Binding (Diderot [1751-65] 1965c)

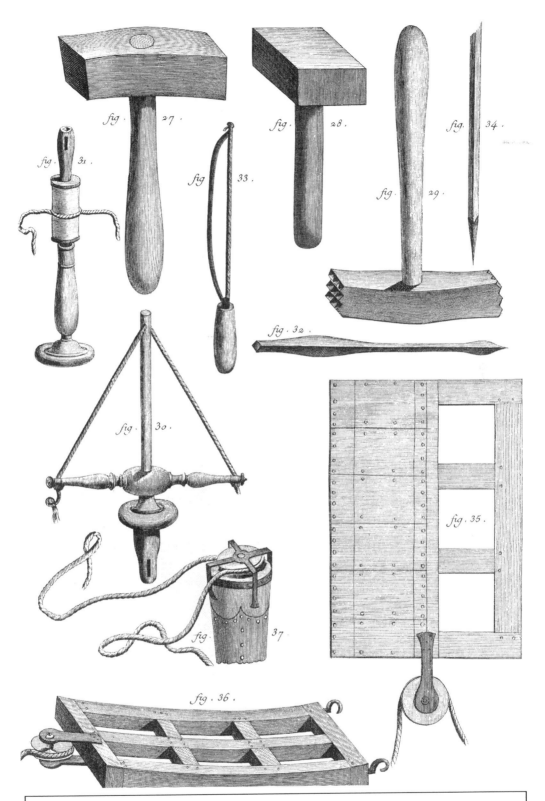

Figure 775. Marble Sculpture (Diderot [1751-65] 1965c)

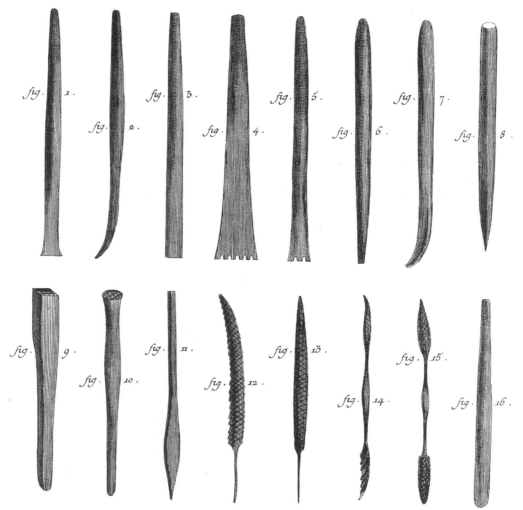

Figure 776. Marble Sculpture (Diderot [1751-65] 1965c)

178

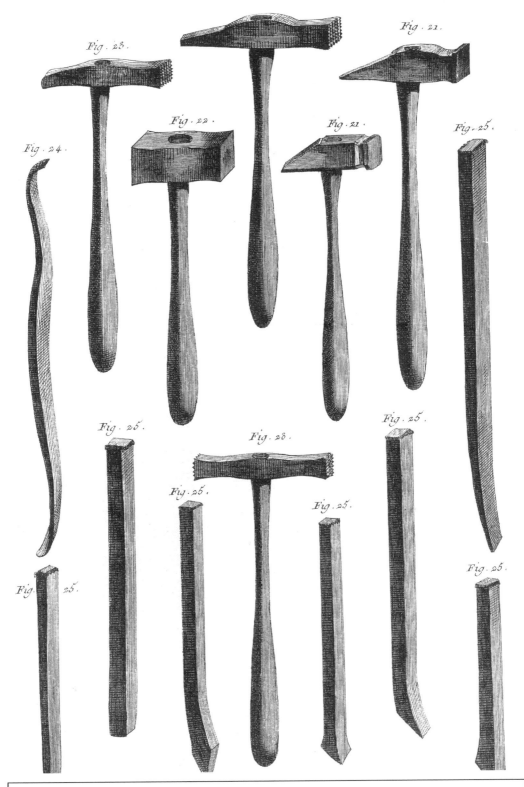

Figure 777. Lead Sculpture (Diderot [1751-65] 1965d)

179

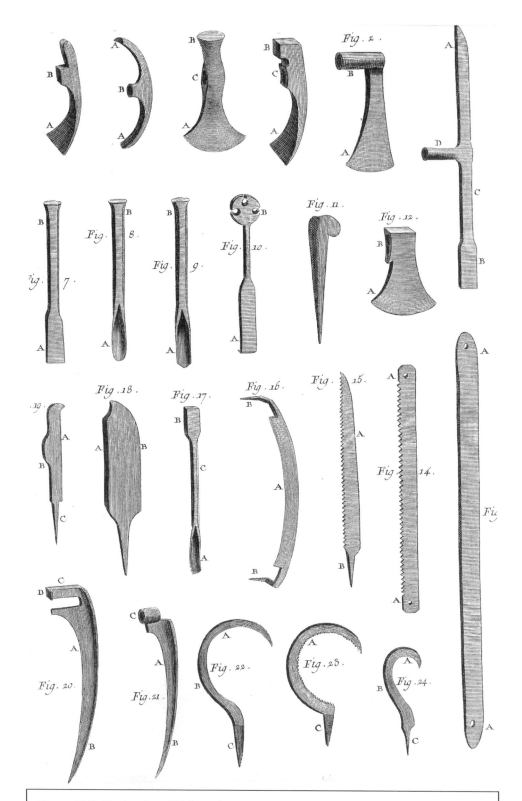

Figure 778. Toolmaker, Whitesmith (Diderot [1751-65] 1965d)

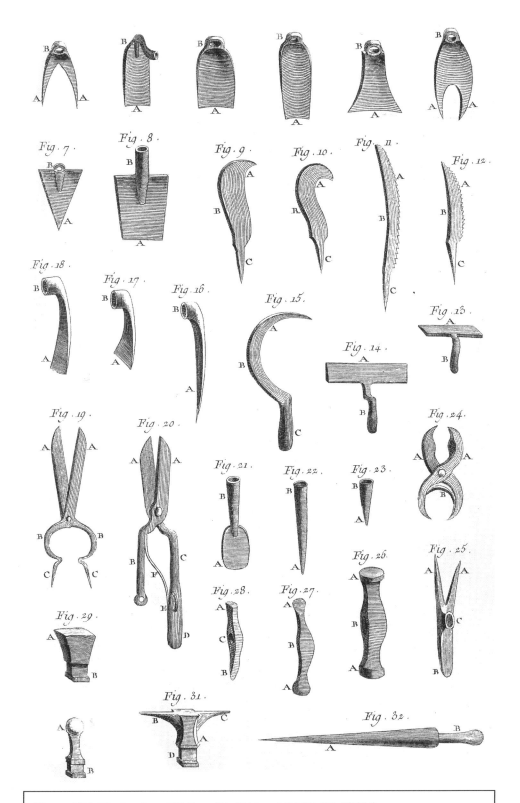

Figure 779. Toolmaker, Whitesmith (Diderot [1751-65] 1965d)

181

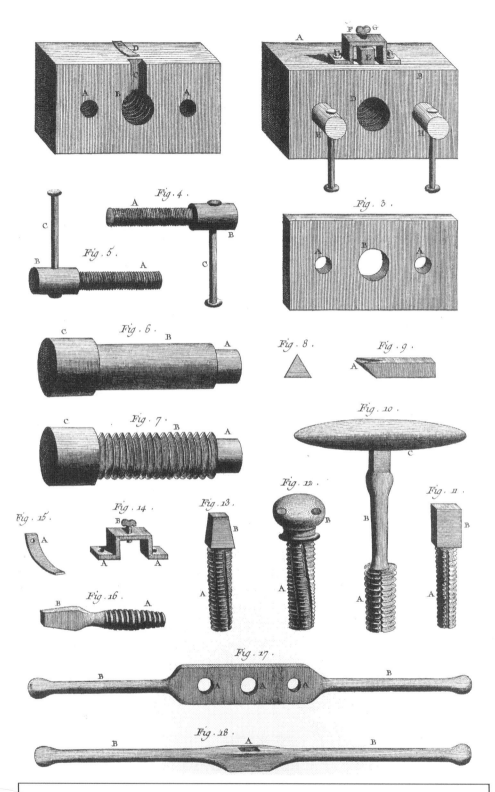

Figure 780. Turning Taps and Dies (Diderot [1751-65] 1965d)

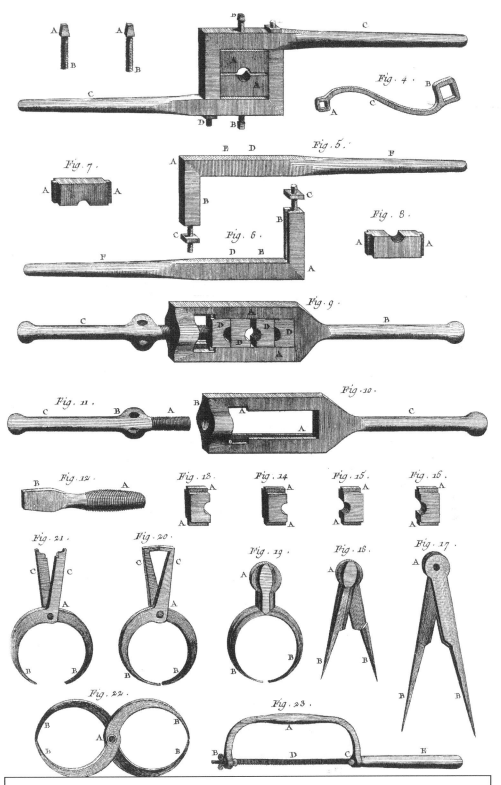

Figure 781. Turning Taps, Dies, and Compasses (Diderot [1751-65] 1965d)

183

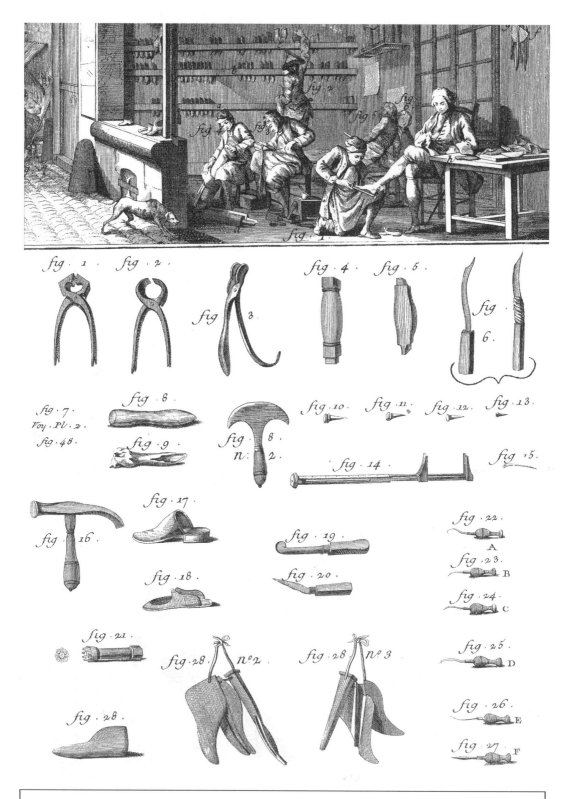

Figure 782. Cobbler (Diderot [1751-65] 1965a)

184

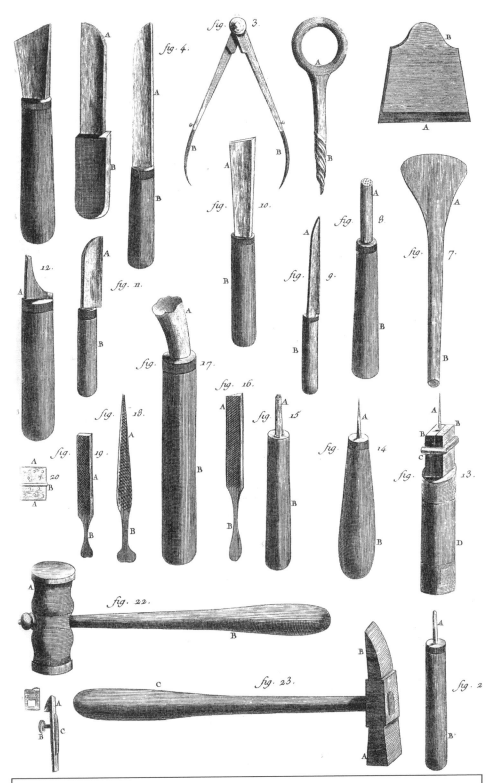

Figure 783. Leather Corsets (Diderot [1751-65] 1965a)

185

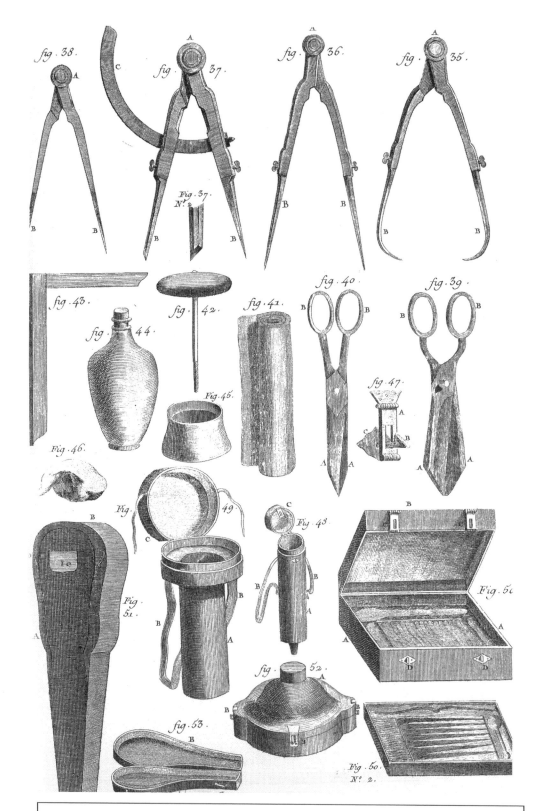

Figure 784. Leather Corsets (Diderot [1751-65] 1965a)

186

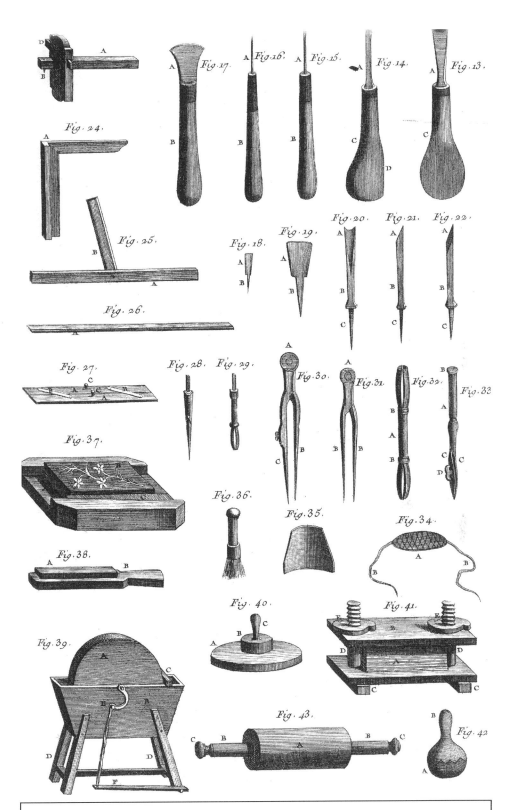

Figure 785. Wood Engraving Carpentry (Diderot [1751-65] 1965b)

187

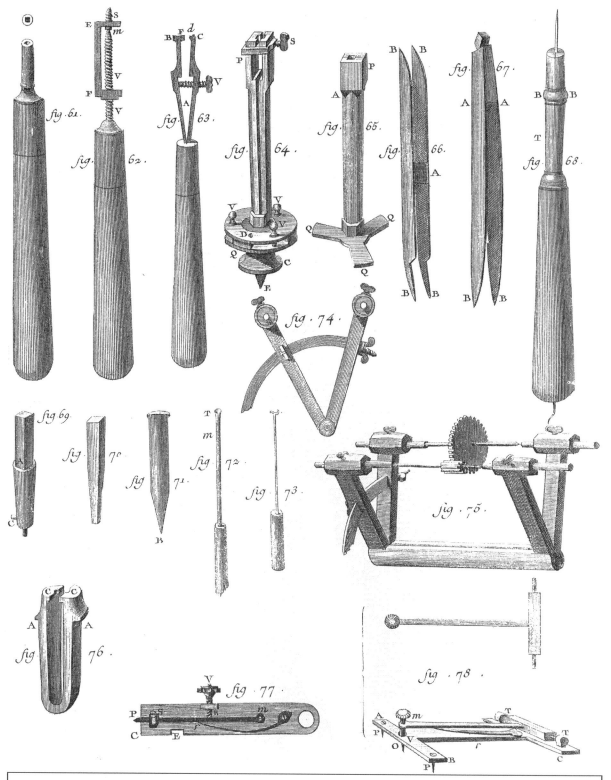

Figure 786. Clockmaker (Diderot [1751-65] 1965a)

188

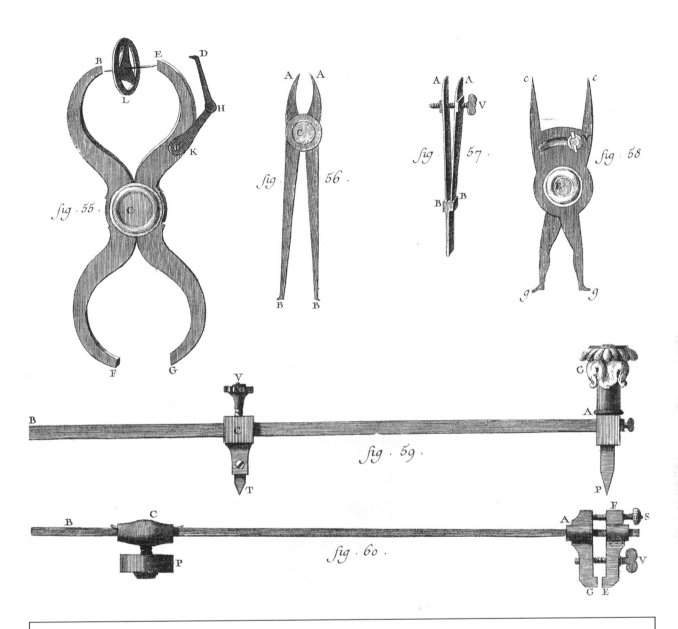

Figure 787. Clockmaker (Diderot [1751-65] 1965a)

189

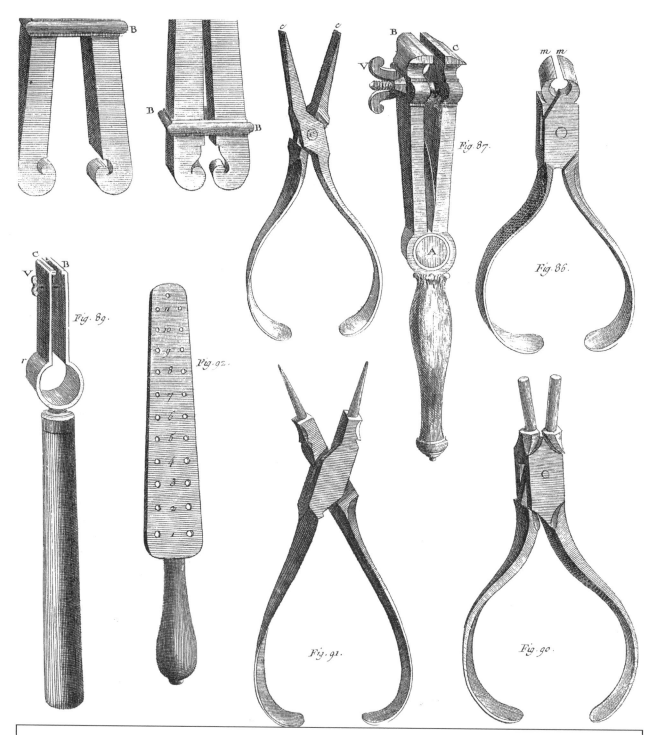

Figure 788. Clockmaker (Diderot [1751-65] 1965a)

190

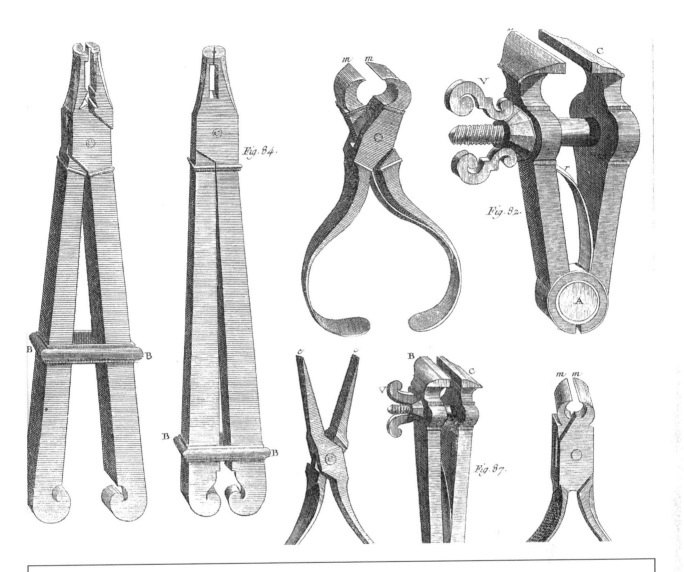

Figure 789. Clockmaker (Diderot [1751-65] 1965a)

191

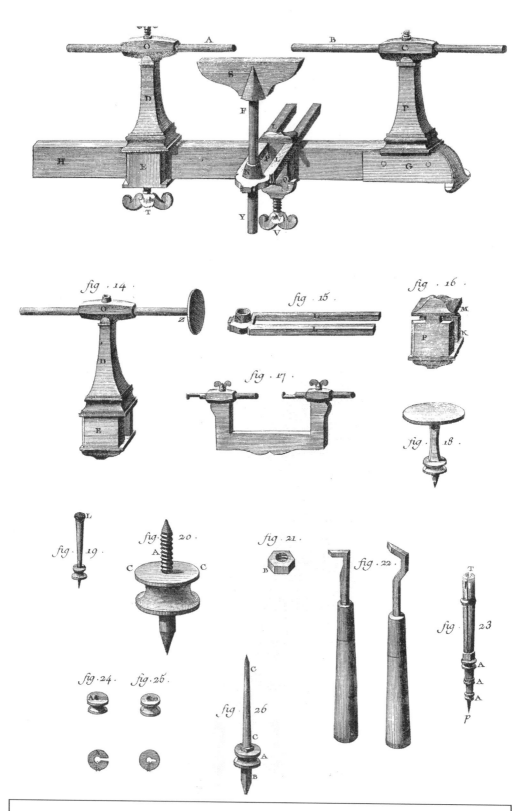

Figure 790. Clockmaker (Diderot [1751-65] 1965a)

192

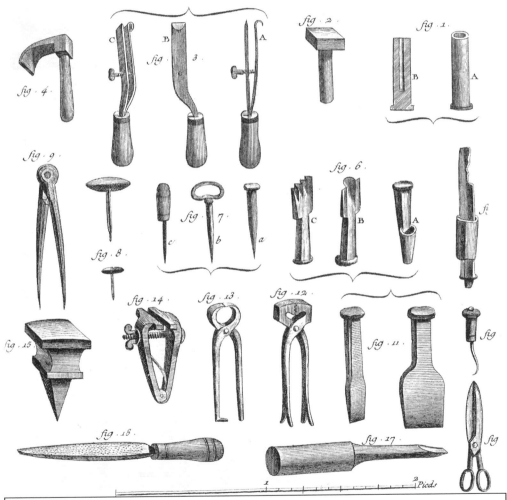

Figure 791. Chest Maker (Diderot [1751-65] 1965a)

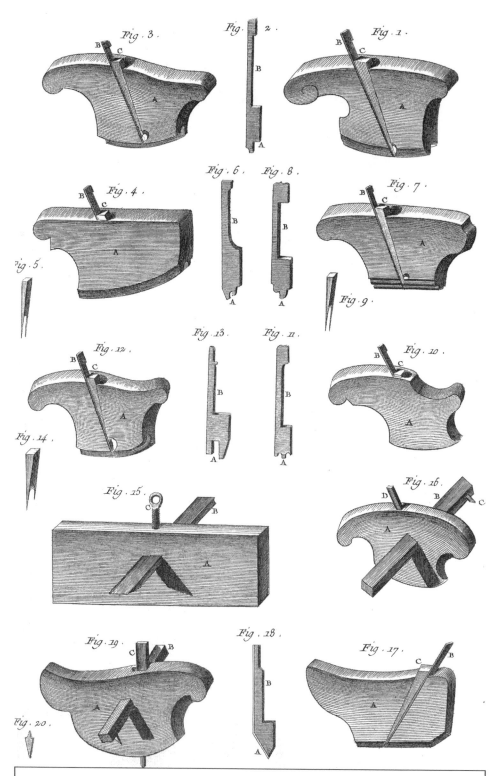

Figure 792. Carpentry, Joining (Diderot [1751-65] 1965c)

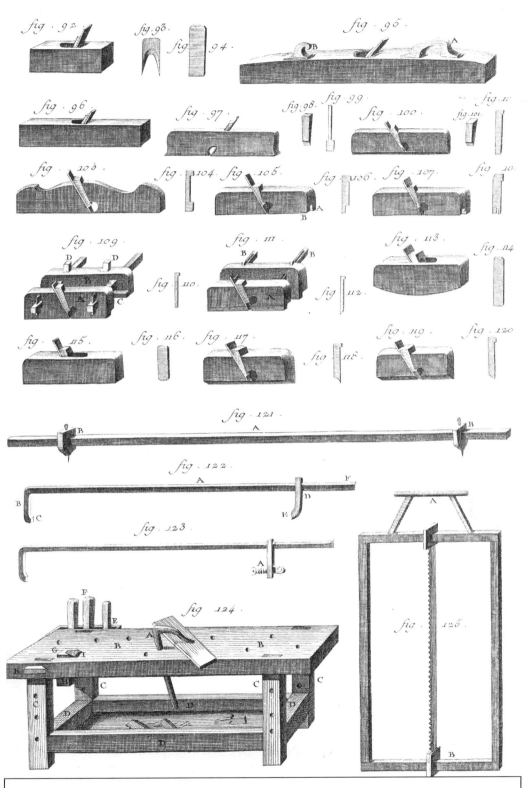

Figure 793. Carpentry, Joining (Diderot [1751-65] 1965c)

195

Figure 794. Whitesmith, Chasing (Diderot [1751-65] 1965a)

196

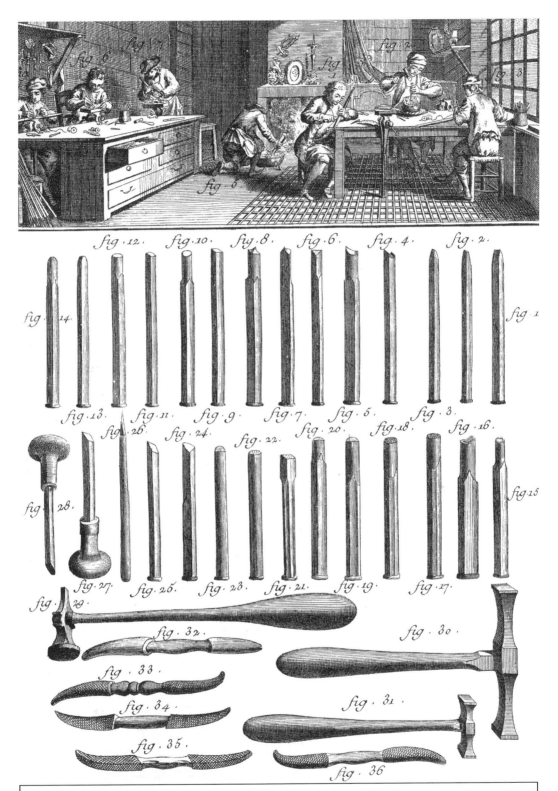

Figure 795. Whitesmith, Chasing (Diderot [1751-65] 1965a)

197

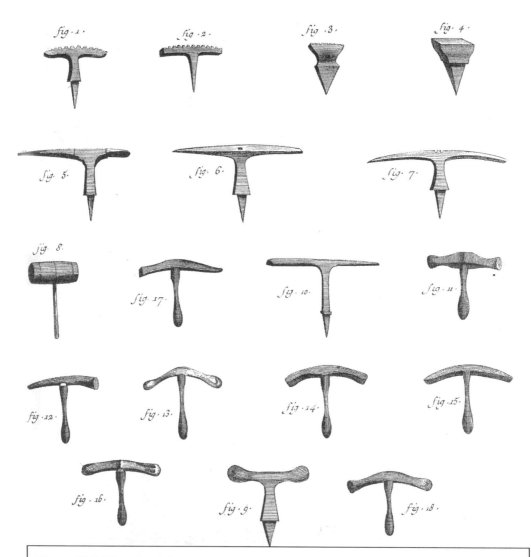

Figure 796. Tinsmith (Diderot [1751-65] 1965a)

198

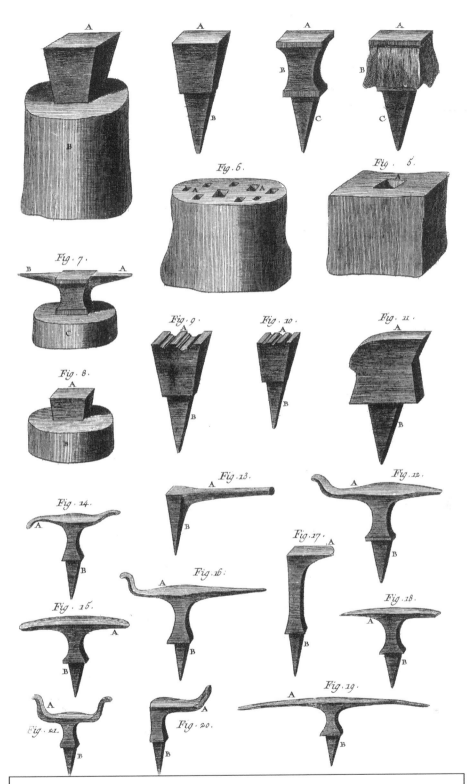

Figure 797. Gold and Silversmith (Diderot [1751-65] 1965c)

199

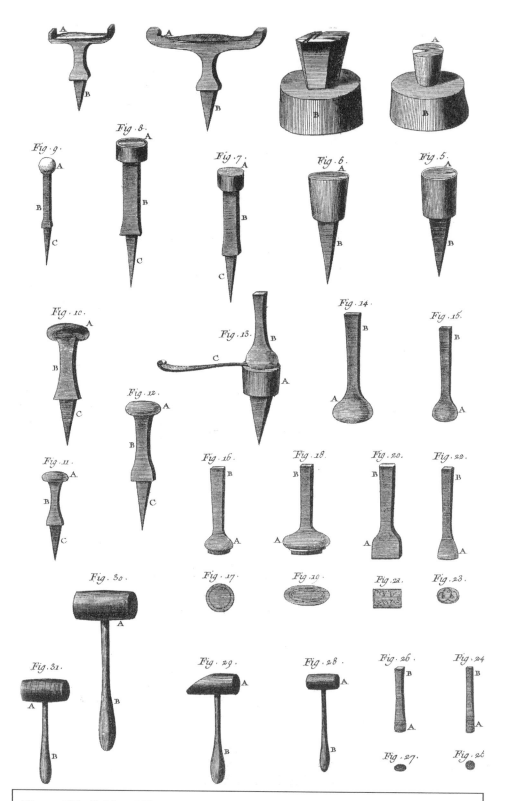

Figure 798. Gold and Silver Smith (Diderot [1751-65] 1965c)

200

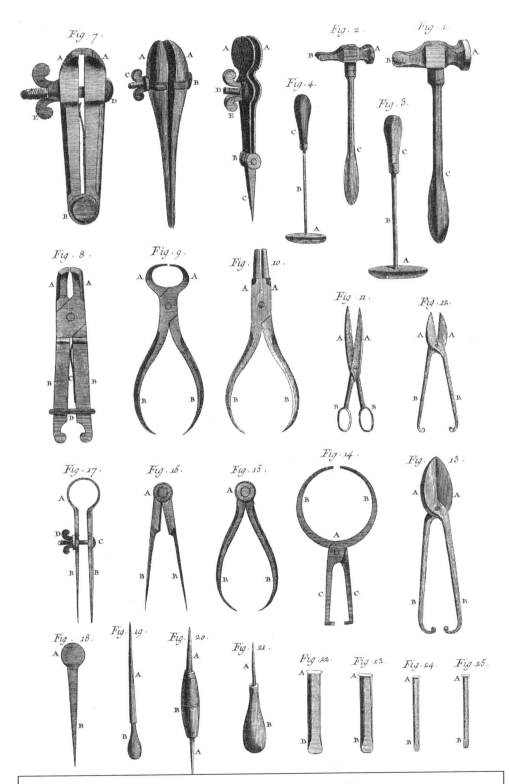

Figure 799. Gold and Silver Smith (Diderot [1751-65] 1965c)

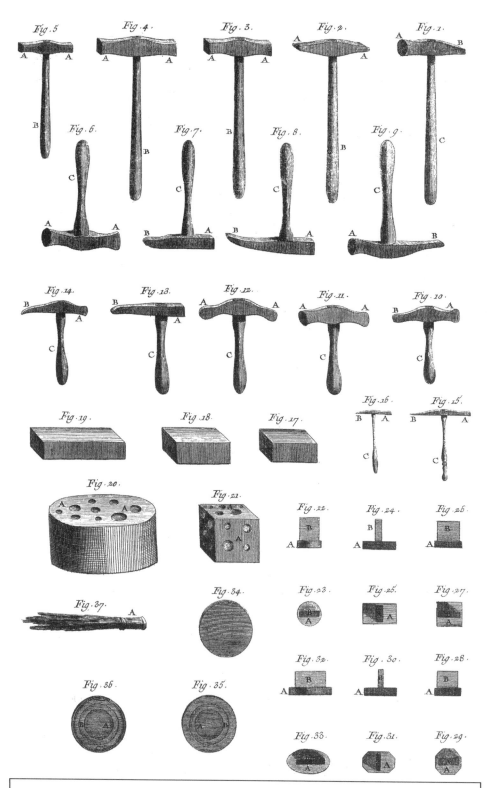

Figure 800. Gold and Silver Smith (Diderot [1751-65] 1965c)

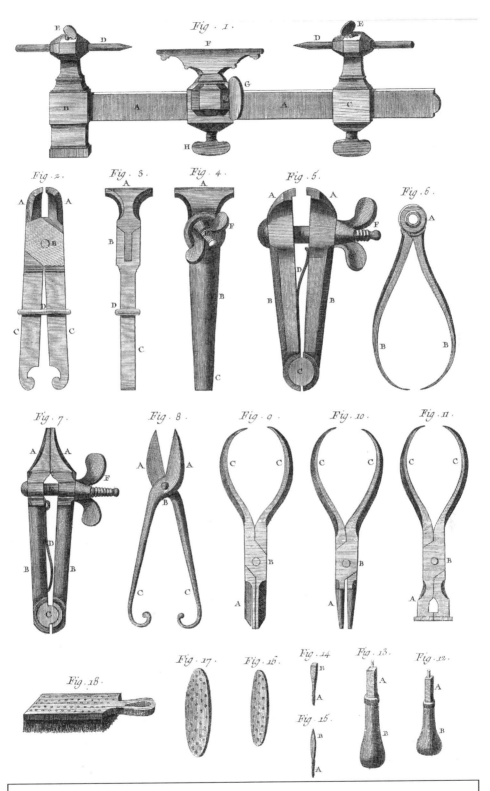

Figure 801. Gold and Silver Smith (Diderot [1751-65] 1965c)

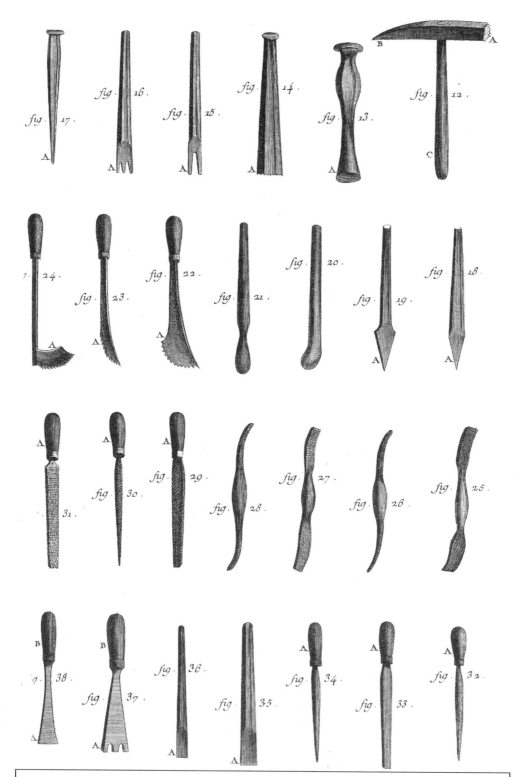

Figure 802. Masonry, Marble (Diderot [1751-65] 1965b)

204

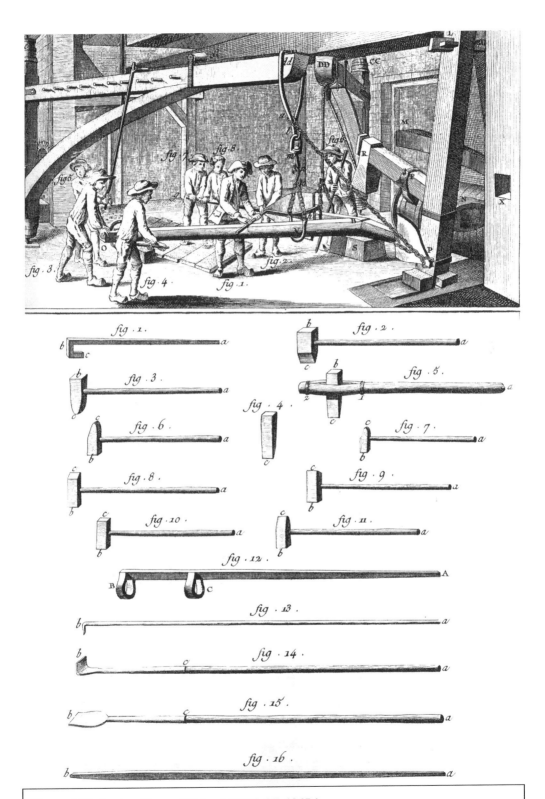

Figure 803. Anchor Forge (Diderot [1751-65] 1965c)

205

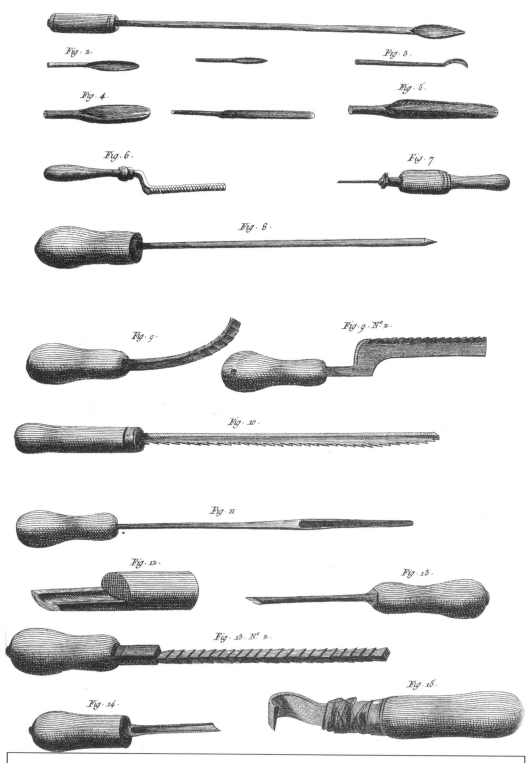

Figure 804. Luthier (Diderot [1751-65] 1965b)

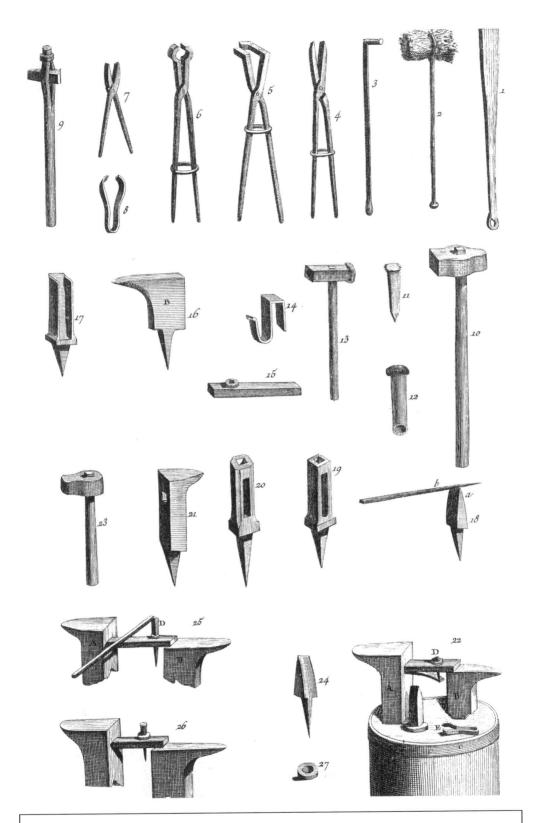

Figure 805. Nail Maker (Diderot [1751-65] 1965a)

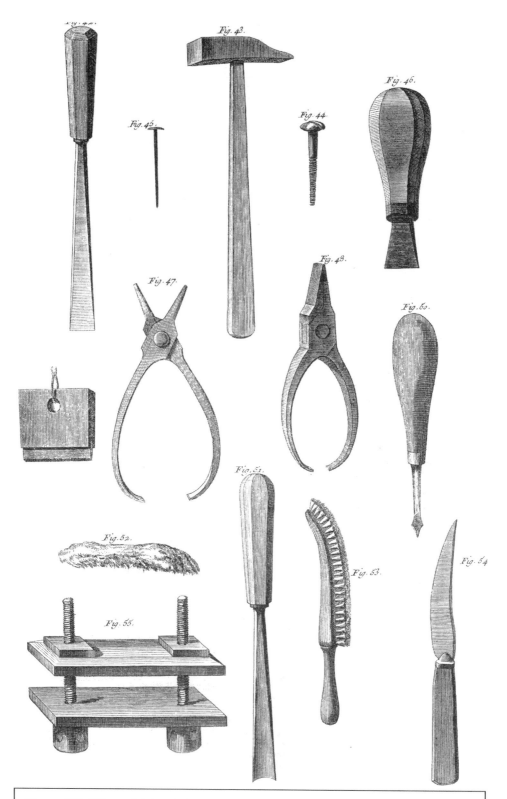

Figure 806. Mirror Maker (Diderot [1751-65] 1965c)

208

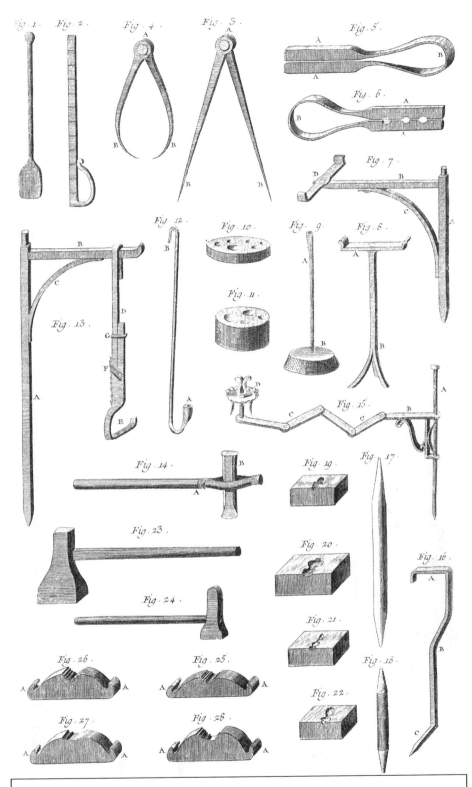

Figure 807. Locksmith (Diderot [1751-65] 1965d)

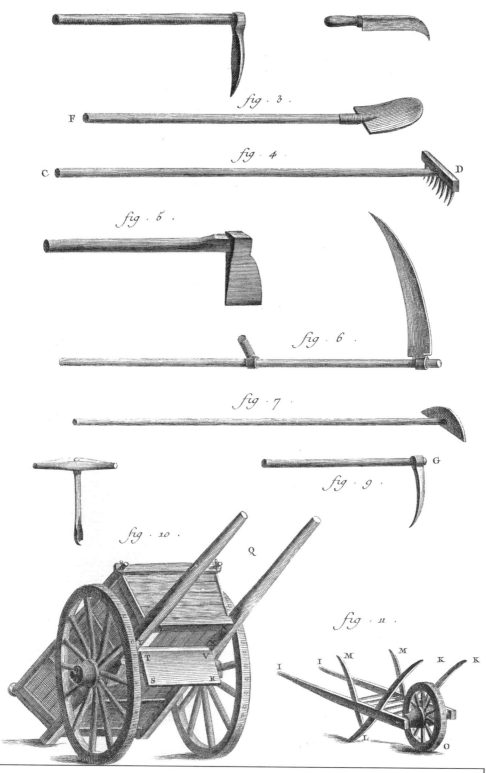

Figure 808. Farm Tools (Diderot [1751-65] 1964)

Appendix 2: Roger Majorowicz Ax Collection

Roger has been a tool collector for much of his adult life, sometimes incorporating tools into his sculpture. In many cases the identity, function, and shape of tools within Roger's sculpture are hidden, becoming subservient components of a larger composition that makes no direct commentary on the phenomenology of tools.

Most of the axes illustrated in this appendix, which include annotations describing or identifying each ax, were collected by Roger after moving to Maine in 1983. The collection constitutes an important contribution to this *Tools Teach* publication, allowing readers of all ages to take a journey through the iconography of American axes. These axes were among the prime movers of the Age of Wind and Wood, helping to build a nation that soon manufactured machine-made axes, which are also a component of this collection.

Biography

Originally from South Dakota, Roger Majorowicz graduated from the Minneapolis College of Art and was awarded a full scholarship to attend the Skowhegan School of Painting and Sculpture in Maine, where he won the school's top award in sculpture. He received a Fulbright Fellowship to study for two years at the Institute D'Arte in Florence, Italy.

Prior to his permanent move to Maine in 1983, he spent over two decades teaching sculpture at the Maryland Institute of Art in Baltimore, with summer retreats to Whitefield, Maine, where he currently resides and works. His work has been featured in museums and galleries around the world, including Carerra, Milan, Spoleto, and Rome, Italy; Munich, Germany; and in the United States in New York, Chicago, Los Angeles, Washington, D.C., and many other cities. Majorowicz has created over 40 public commissions around the United States, most of them monumental in scale. Many of his sculptures can be seen in schools and public buildings in Maine, commissioned by the Maine Arts Commission Percent for Art program, including Leavitt Area High School in Turner, Westbrook High School, The University of Maine at Augusta, and Whitefield Elementary School.

Majorowicz's home and studio along the Sheepscot River borders a field enhanced by his wildly fantastic sculptures, some over 30 feet tall with moveable parts. He draws inspiration from many sources, including mythical themes, the human figure, and the landscape.

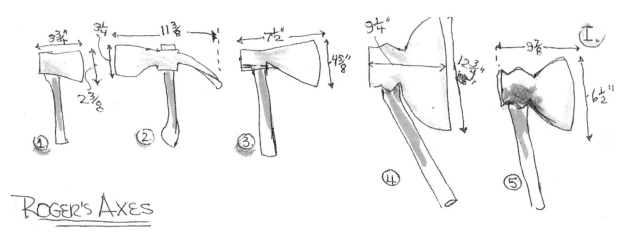

ROGER'S AXES

TOP ROW IN STUDIO
1. MY SMALLEST AXE (NO MARKS)
2. AXE-CHISEL —FORGED FROM AN ADZE BY MY FATHER — R.W. MAJOROWICZ
   IN TIMBER LAKE, S.D. — 1961 (GREEN PAINT — M on handle
3. ITALIAN AXE — PURCHASED AT FLEA-MARKET, FLORENCE, ITALY, 1959 (24¢)
   MARKED   (GREEN PAINT.)
         000
4. PENNSYLVANIA AXE — PURCHASED FROM GRADUATE STUDENT JOHN
              (1968)
   WILLIAMS — MD. INST. of ART. (ORIGINALLY FROM LANCASTER, P.A.
   THIS IS MY LARGEST BROAD AXE, — 12 3/4" BLADE — NO MARKS
   GREAT SHAPE — ORIGINAL HANDLE
5. RUSTED, VERY OLD — NO MARKS.

212

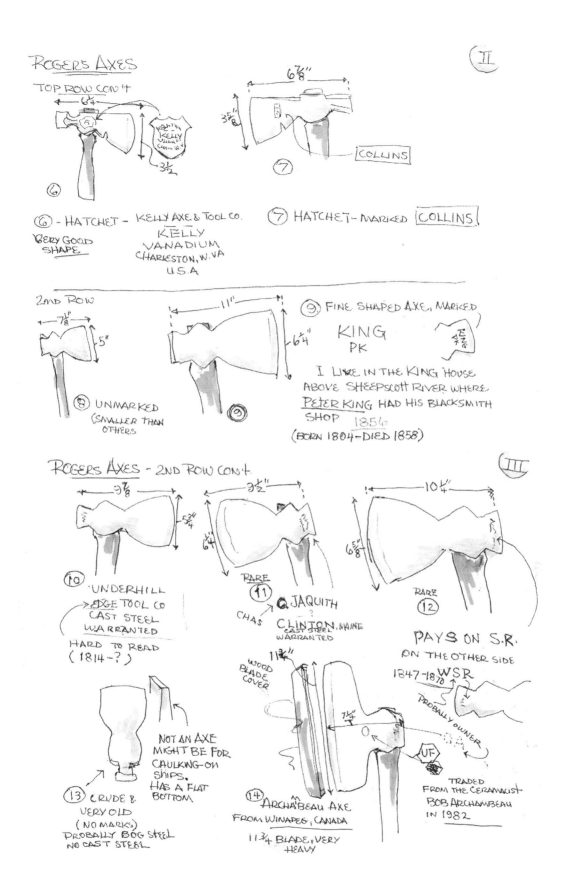

ROGERS AXES

TOP ROW CON'T

6¼"

3½"

COLLINS

6⅞"

3¾"

⑥

⑦

⑥ - HATCHET - KELLY AXE & TOOL CO.
KELLY
VANADIUM
CHARLESTON, W. VA
U.S.A
VERY GOOD
SHAPE

⑦ HATCHET - MARKED COLLINS

2ND ROW

7⅛"

5"

11"

6¼"

⑧ UNMARKED
(SMALLER THAN
OTHERS

⑨

⑨ FINE SHAPED AXE, MARKED
KING
PK
KING
PK

I LIVE IN THE KING HOUSE
ABOVE SHEEPSCOTT RIVER WHERE
PETER KING HAD HIS BLACKSMITH
SHOP 1856
(BORN 1804-DIED 1858)

ROGERS AXES - 2ND ROW CON'T

3⅞"

5¾"

3½"

6¼"

10¼"

6⅝"

⑩ 'UNDERHILL
EDGE TOOL CO
CAST STEEL
WARRANTED
HARD TO READ
(1814-?)

RARE
⑪
Q. JAQUITH
CHAS
CLINTON, MAINE
CAST STEEL
WARRANTED

RARE
⑫

PAYS ON S.R.
ON THE OTHER SIDE
1847-18⁷⁰ WSR
PROBALLY OWNER

⑬ CRUDE &
VERY OLD
(NO MARKS)
PROBALLY BOG STEEL
NO CAST STEEL

NOT AN AXE
MIGHT BE FOR
CAULKING - ON
SHIPS.
HAS A FLAT
BOTTOM

WOOD
BLADE
COVER

11¼"

7¼"

UF

⑭ ARCHA'BEAU AXE
FROM WINAPEG, CANADA
11¾ BLADE, VERY
HEAVY

TRADED
FROM THE CERAMACIST
BOB ARCHAMBEAU
IN 1982

213

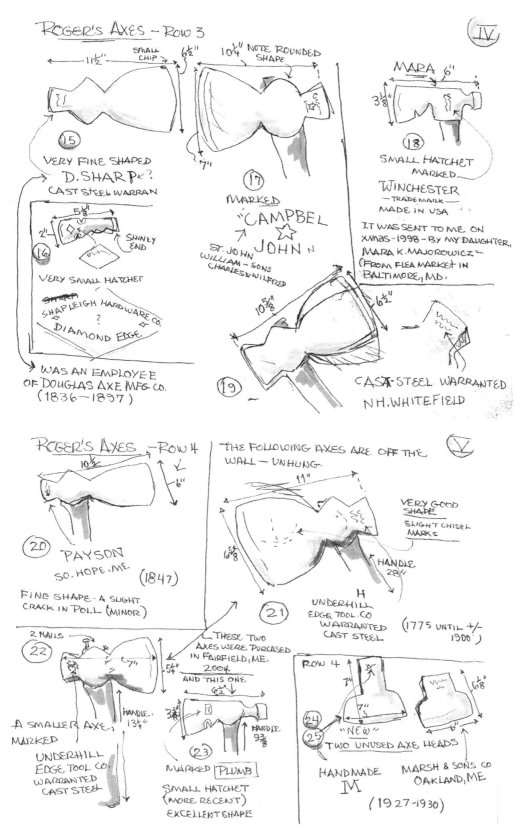

ROGER'S AXES - ROW 3   ④ IV

SMALL CHIP → 6½"   10¼" NOTE ROUNDED SHAPE

11½"

⑮
VERY FINE SHAPED
→ D. SHARPE?
CAST STEEL WARRAN

7"

⑰
MARKED
"CAMPBEL ☆
JOHN N
ST. JOHN
WILLIAM - SONS
CHARLES & WILFRED

MARA 6"
3⅛"

⑱
SMALL HATCHET
MARKED
WINCHESTER
— TRADEMARK —
MADE IN USA

IT WAS SENT TO ME ON
XMAS-1998 - BY MY DAUGHTER,
MARA K. MAJOROWICZ
(FROM FLEA MARKET IN
BALTIMORE, MD.

5⅞"
2"
⑯
SHINEY END

VERY SMALL HATCHET
SHARP
SHAPLEIGH HARDWARE CO.
?
DIAMOND EDGE

→ WAS AN EMPLOYEE
OF DOUGLAS AXE MFG. CO.
(1836 - 1897)

10⅝"   6½"

⑲
CAST-STEEL WARRANTED
N.H. WHITEFIELD

ROGER'S AXES - ROW 4   THE FOLLOWING AXES ARE OFF THE
                           WALL — UNHUNG.   ⑤ V

10½"   6"

⑳
PAYSON
SO. HOPE. ME.   (1847)

FINE SHAPE - A SLIGHT
CRACK IN POLL (MINOR)

11"
VERY GOOD
SHAPE
SLIGHT CHISEL
MARKS

5⅞"
HANDLE
28"

㉑
H
UNDERHILL
EDGE TOOL CO
WARRANTED
CAST STEEL   (1775 UNTIL +/-
                      1900')

2 NAILS
㉒
7"
5¼"

A SMALLER AXE,
MARKED
UNDERHILL
EDGE TOOL CO
WARRANTED
CAST STEEL

HANDLE.
13¼"

THESE TWO
AXES WERE PURCASED
IN FAIRFIELD, ME.
2004
AND THIS ONE
6½"

3¾"
㉓
HANDLE
9⅜"
MARKED [PLUMB]
SMALL HATCHET
(MORE RECENT)
EXCELLENT SHAPE

ROW 4
7"
7"   6⅛"
㉔
㉕   6"
"NEW"
TWO UNUSED AXE HEADS
HANDMADE   MARSH & SONS CO
M   OAKLAND, ME
(1927-1930)

214

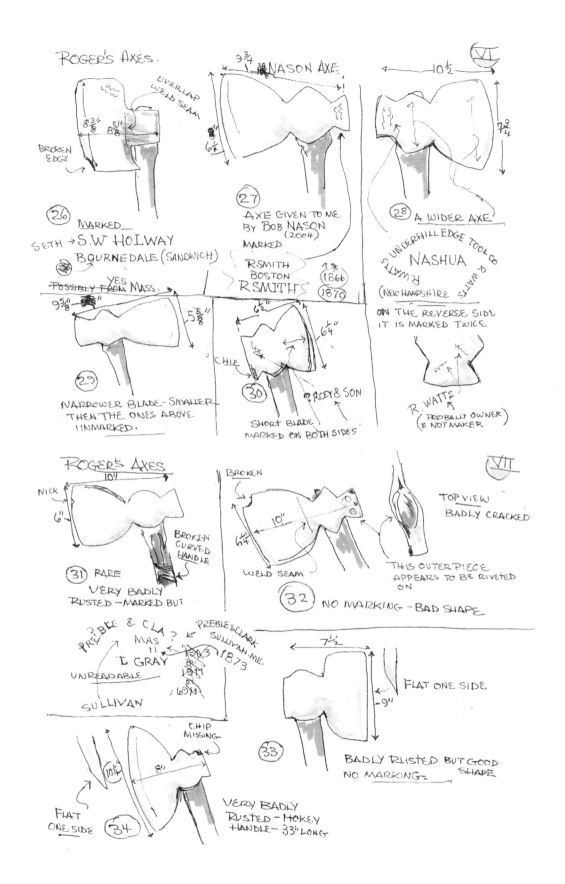

ROGER'S AXES.

OVERLAP WELD SEAM

8⅜" 5"/8"
BROKEN EDGE

㉖ MARKED
SETH → S.W HOLWAY
BOURNEDALE (SANDWICH)
POSSIBLY FROM MASS.

9⅜"
5⅝/100"

㉙ NARROWER BLADE - SMALLER
THEN THE ONES ABOVE.
UNMARKED.

9¾" NASON AXE

6¼"

㉗ AXE GIVEN TO ME
BY BOB NASON
(2004)
MARKED
R.SMITH
BOSTON
R.SMITHS

1866
1870

6½"
6¼"
CHIP

㉚ SHORT BLADE
MARKED ON BOTH SIDES
PERCEY & SON

VI

10½"

7¾/4"

㉘ A WIDER AXE
UNDERHILL EDGE TOOL C⁰
R. WATTS
NASHUA
(NEW HAMPSHIRE)
ON THE REVERSE SIDE
IT IS MARKED TWICE.

R WATTS
(PROBABLY OWNER
& NOT MAKER)

ROGER'S AXES
10"

NICK
6"

㉛ RARE
VERY BADLY
RUSTED - MARKED BUT
BROKEN CURVED HANDLE

BROKEN
10"
6¼"
WELD SEAM

㉜ NO MARKING - BAD SHAPE

TOP VIEW
BADLY CRACKED

THIS OUTER PIECE
APPEARS TO BE RIVETED ON

VII

PREBLE & CLA?
PREBLE & CLARK
SULLIVAN·ME.
MAS?
L GRAY
1863 1873
UNREADABLE
19M
SULLIVAN
16M

CHIP MISSING
10½"
8"

FLAT ONE SIDE

㉞ VERY BADLY
RUSTED - HOKEY
HANDLE - 33" LONG

7½"

FLAT ONE SIDE
9"

㉝ BADLY RUSTED BUT GOOD SHAPE
NO MARKING·

215

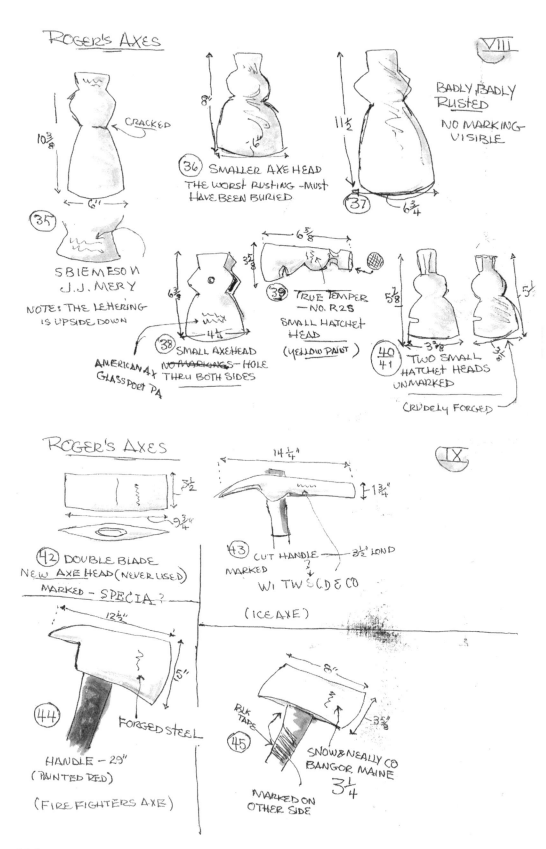

ROGER'S AXES

VIII

CRACKED

$10\frac{3}{8}$

6"

35

SBIEMESON
J.J. MERY

NOTE: THE LETTERING
IS UPSIDE DOWN

8"

36 SMALLER AXE HEAD
THE WORST RUSTING - MUST
HAVE BEEN BURIED

$11\frac{1}{2}$

BADLY, BADLY
RUSTED

NO MARKING
VISIBLE

37 $6\frac{3}{4}$

$6\frac{5}{8}$

$3\frac{1}{2}$

39 TRUE TEMPER
— NO. R25

SMALL HATCHET
HEAD

(YELLOW PAINT)

$6\frac{3}{8}$

38 SMALL AXE HEAD
NO MARKINGS - HOLE
THRU BOTH SIDES

$4\frac{1}{4}$

AMERICAN AX
GLASSPORT PA

$5\frac{7}{8}$

$5\frac{1}{2}$

40
41 TWO SMALL
HATCHET HEADS
UNMARKED

$3\frac{3}{8}$

CRUDELY FORGED

ROGER'S AXES

IX

$3\frac{1}{2}$

$9\frac{3}{4}$

42 DOUBLE BLADE
NEW AXE HEAD (NEVER USED)

MARKED - SPECIA ?

$14\frac{1}{4}$"

$1\frac{3}{4}$"

43 CUT HANDLE — $3\frac{1}{2}$" LONG
MARKED

WI TW&CD&CO

(ICE AXE)

$12\frac{1}{2}$"

5"

44

FORGED STEEL

HANDLE - 29"
(PAINTED RED)

(FIRE FIGHTERS AXE)

8"

BLK
TAPE

$3\frac{3}{8}$

45

SNOW & NEALLY CO
BANGOR, MAINE

$3\frac{1}{4}$

MARKED ON
OTHER SIDE

216

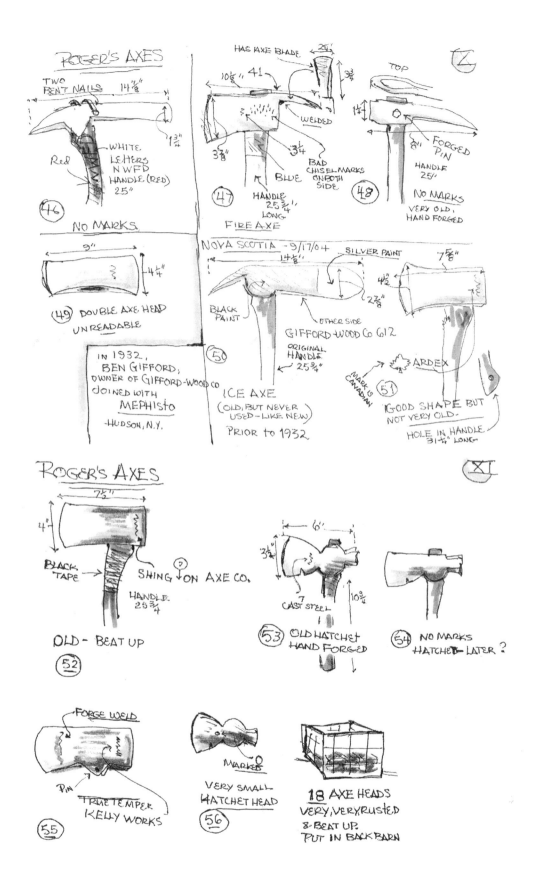

ROGER'S AXES

TWO BENT NAILS   14 7/8"

Red

WHITE LETTERS N W F D HANDLE (RED) 25"

(46)

NO MARKS

HAS AXE BLADE   2 1/4"

10 1/2" 41   3 3/4"

WELDED

3 3/8"   3 1/4"

BLUE   BAD CHISEL MARKS ON BOTH SIDE

HANDLE 25 3/4" LONG

(47) FIRE AXE

TOP

1 1/2"

8" FORGED PIN

HANDLE 25"

(48) NO MARKS VERY OLD, HAND FORGED

3"   4 1/4"

(49) DOUBLE AXE HEAD UNREADABLE

IN 1932, BEN GIFFORD, OWNER OF GIFFORD-WOOD CO JOINED WITH MEPHISTO - HUDSON, N.Y.

NOVA SCOTIA - 9/17/04   14 1/8"   SILVER PAINT   7 5/8"

4 1/2"

2 7/8"

BLACK PAINT

OTHER SIDE

GIFFORD-WOOD CO 612

ORIGINAL HANDLE 25 3/4"

(50) ICE AXE (OLD, BUT NEVER USED - LIKE NEW) PRIOR TO 1932

ARDEX

MARK IS CANADIAN

(51) GOOD SHAPE BUT NOT VERY OLD. HOLE IN HANDLE 31 1/2" LONG

XI

ROGER'S AXES

7 1/2"

4"

BLACK TAPE

SHING?ON AXE CO.

HANDLE 25 3/4"

OLD - BEAT UP

(52)

6"

3 1/2"   10 3/4"

7 CAST STEEL

(53) OLD HATCHET HAND FORGED

(54) NO MARKS HATCHET - LATER ?

FORGE WELD

PIN

TRUE TEMPER KELLY WORKS

(55)

MARKED

VERY SMALL HATCHET HEAD

(56)

18 AXE HEADS VERY, VERY, RUSTED & BEAT UP. PUT IN BACK BARN

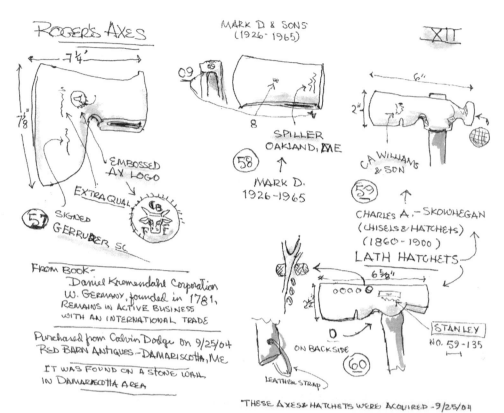

ROGER'S AXES

MARK D & SONS
(1926-1965)

XII

7¼"

7⅞"

EMBOSSED
AX LOGO

EXTRA QUAL

⑤⑦ SIGNED
GERRUBER, SC

09

8

SPILLER
OAKLAND, ME

⑤⑧

MARK D.
1926-1965

6"

2"

CA WILLIAMS
& SON

⑤⑨

CHARLES A. - SKOWHEGAN
(CHISELS & HATCHETS)
(1860-1900)

LATH HATCHETS

6⅝"

STANLEY
NO. 59-135

ON BACKSIDE

⑥⓪

LEATHER STRAP

FROM BOOK -
   Daniel Kremendahl Corporation
   W. Germany, founded in 1781,
   REMAINS IN ACTIVE BUSINESS
   WITH AN INTERNATIONAL TRADE

Purchased from Calvin Dodge on 9/25/04
Red Barn Antiques - Damariscotta, Me

IT WAS FOUND ON A STONE WALL
IN DAMARISCOTTA AREA

THESE AXES & HATCHETS WERE ACQUIRED - 9/25/04

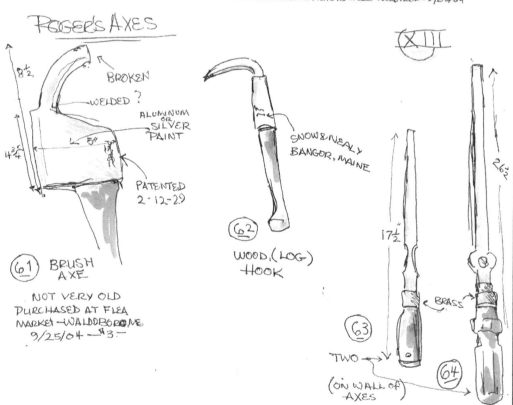

ROGER'S AXES

XIII

8½"

BROKEN

WELDED ?

ALUMINUM
OR
SILVER
PAINT

4¾"

5"

PATENTED
2-12-29

⑥① BRUSH
AXE

NOT VERY OLD
PURCHASED AT FLEA
MARKET - WALDOBORO ME
9/25/04 - $3 -

SNOW & NEALY
BANGOR, MAINE

⑥②

WOOD, (LOG)
HOOK

26½"

17½"

BRASS

⑥③

⑥④

TWO

(ON WALL OF)
AXES

218

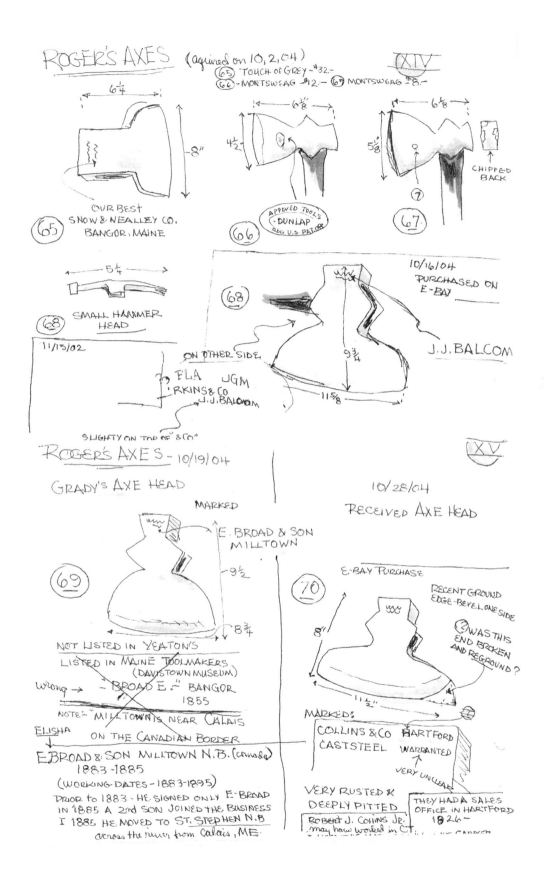

ROGER'S AXES (aquired on 10, 2, 04)
(65) TOUCH OF GREY - $32.-
(66) MONTSWEAG - $12.- (67) MONTSWAG - $8.-

XIV

6¼"
8"

OUR BEST
(65) SNOW & NEALLEY CO.
BANGOR, MAINE

6⅛"
4½"

(66) APPROVED TOOLS
DUNLAP
REG. U.S. PAT. OFF.

6⅛"
5⅛"

CHIPPED BACK

(7)

(67)

5¼"
(68) SMALL HAMMER HEAD

11/15/02

(68)

10/16/04
PURCHASED ON
E-BAY

9¾"
11⅝"

J.J. BALCOM

ON OTHER SIDE
FLA   JGM
RKINS & CO
J.J. BALCOM

SLIGHTY ON TOP OF "& CO"

ROGER'S AXES - 10/19/04

GRADY'S AXE HEAD

MARKED
E. BROAD & SON
MILLTOWN

(69)

9½
8¾

NOT LISTED IN YEATON'S
LISTED IN MAINE TOOLMAKERS
(DAVISTOWN MUSEUM)
Wrong → BROAD E. - BANGOR
1855
NOTE - MILLTOWN IS NEAR CALAIS
ELISHA   ON THE CANADIAN BORDER
E. BROAD & SON MILLTOWN N.B. (canada)
1883-1885
(WORKING DATES - 1883-1895)
PRIOR TO 1883 - HE SIGNED ONLY E. BROAD
IN 1885 A 2nd SON JOINED THE BUSINESS
I 1885 HE MOVED TO ST. STEPHEN N.B
across the river from Calais, ME.

XV

10/28/04

RECEIVED AXE HEAD

E-BAY PURCHASE

RECENT GROUND
EDGE-BEVEL ONE SIDE

(?) WAS THIS
END BROKEN
AND REGROUND?

(70)

8"
11½"

MARKED:
COLLINS & CO   HARTFORD
CAST STEEL   WARRANTED
VERY UNCLEAR

VERY RUSTED &
DEEPLY PITTED

THEY HAD A SALES
OFFICE IN HARTFORD
1826-

Robert J. Collins Jr.
may have worked in CT

219

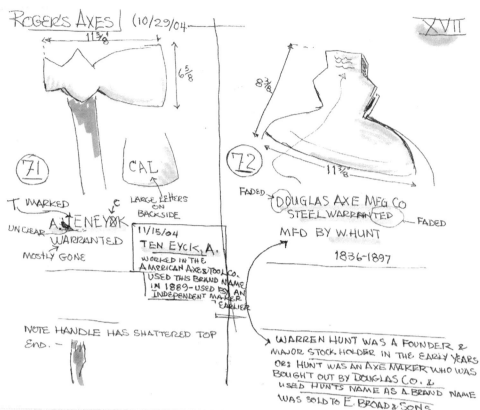

ROGER'S AXES (10/29/04)

XVII

11 3/8"

6 5/8"

8 3/8"

71

CAL

11 3/8"

T. MARKED

LARGE LETTERS
ON
BACKSIDE

A. TENEYØK

FADED → DOUGLAS AXE MFG CO
STEEL WARRANTED ← FADED

UNCLEAR → WARRANTED

MOSTLY GONE

72

MFD BY W. HUNT

11/15/04
TEN EYCK, A.
WORKED IN THE
AMERICAN AXE & TOOL CO.
USED THIS BRAND NAME
IN 1889 - USED BY AN
INDEPENDENT MAKER
EARLIER

1836-1897

NOTE HANDLE HAS SHATTERED TOP
END. —

WARREN HUNT WAS A FOUNDER &
MAJOR STOCK HOLDER IN THE EARLY YEARS
OR: HUNT WAS AN AXE MAKER WHO WAS
BOUGHT OUT BY DOUGLAS CO. &
USED HUNTS NAME AS A BRAND NAME
WAS SOLD TO E. BROAD & SONS

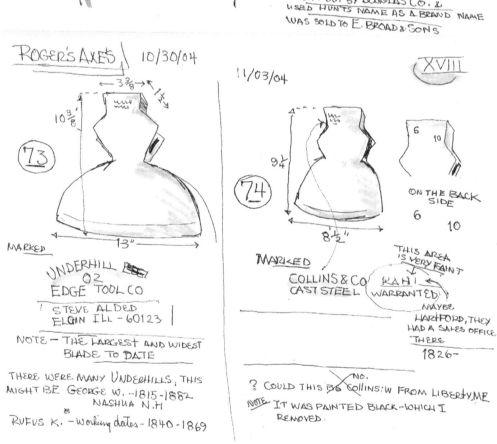

ROGER'S AXES 10/30/04

XVIII

11/03/04

3 3/8"

1 3/8"

10 3/8"

73

9 1/4"

6   10

74

ON THE BACK
SIDE

6   10

13"

MARKED

UNDERHILL
O2
EDGE TOOL CO

8 1/2"

MARKED

THIS AREA
IS VERY FAINT

STEVE ALDED
ELGIN ILL - 60123

COLLINS & CO
CAST STEEL

KAH1
WARRANTED ?

NOTE - THE LARGEST AND WIDEST
BLADE TO DATE

MAYBE
HARTFORD, THEY
HAD A SALES OFFICE
THERE

1826-

THERE WERE MANY UNDERHILLS, THIS
MIGHT BE GEORGE W. -1815-1882
NASHUA N.H
OR
RUFUS K. - Working dates - 1840-1869

? COULD THIS BE COLLINS:W FROM LIBERTY,ME   NO.

NOTE IT WAS PAINTED BLACK - WHICH I
REMOVED.

220

ROGER's AXES (11/04/04)

OLD USA

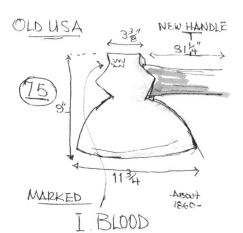

(75)

NEW HANDLE

3⅜"

31¼"

9"

11¾"

— About 1860 —

MARKED

I. BLOOD

BALLSTON, N.Y.

ISIAH BLOOD - BORN 1810, DIED 1870

GOOD SHAPE - BUT THE HEAD HAS BEEN OVERLY CLEANED.

(11/13/04)

AMOSKEAG AX CO.

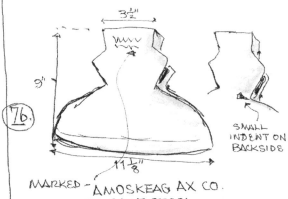

(76)

3½"

9"

11⅛"

SMALL INDENT ON BACKSIDE

MARKED - AMOSKEAG AX CO.
CAST STEEL

MANCHESTER, N.H.
1862-1879

THIS WAS FORMERLY BLODGETT EDGE TOOL MFG. CO., IT WAS ABSORBED BY UNDERHILL EDGE TOOL CO. IN 1879
FROM —

ROGER's AXES (11/13/04

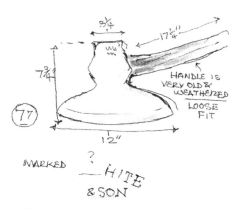

(77)

3¾"

17¼"

7¾"

HANDLE IS VERY OLD & WEATHERED
LOOSE FIT

12"

MARKED ? HITE
& SON

POLL HAS BEEN BEATEN AND SOME BEAT MARKS ON →

11/15/04

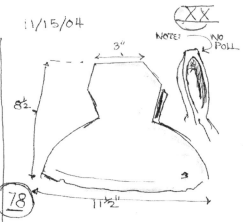

(78)

(XX)

NOTE: NO POLL

3"

8½"

11½"

NO MARKS - NO EVIDENCE OF CAST STEEL - COULD BE FROM BE FROM LATE 1700's OR VERY EARLY 1800's

UNFORTUNATLY IT HAS BEEN PAINTED BLACK

221

ROGER'S AXES

11/15/04

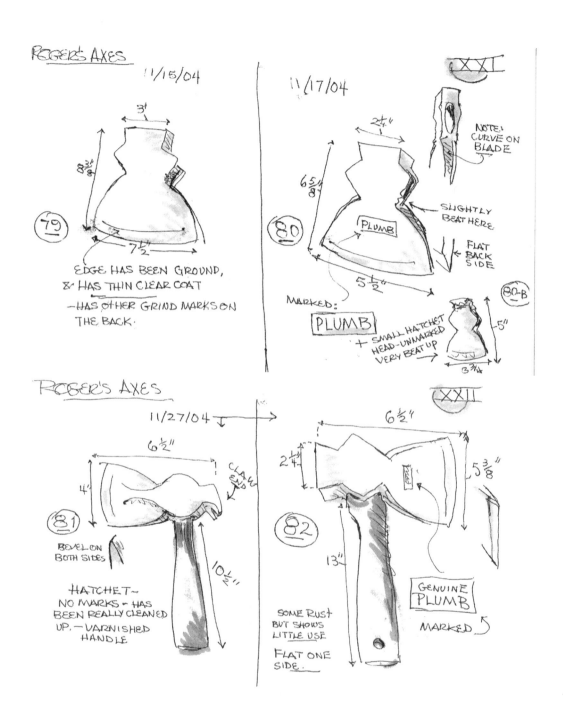

3"

8 3/8

79

7 1/2

EDGE HAS BEEN GROUND,
& HAS THIN CLEAR COAT
— HAS OTHER GRIND MARKS ON
THE BACK.

11/17/04

XXI

NOTE:
CURVE ON
BLADE

2 1/4"

6 5/8

80

PLUMB

SLIGHTLY
BEAT HERE

FLAT
BACK
SIDE

5 1/2"

MARKED:

PLUMB

80-B

+ SMALL HATCHET
HEAD - UNMARKED
VERY BEAT UP

5"

3 3/4"

ROGER'S AXES

11/27/04

6 1/2"

CLAW
END

4"

81

10 1/2"

BEVEL ON
BOTH SIDES

HATCHET ~
NO MARKS - HAS
BEEN REALLY CLEANED
UP. — VARNISHED
HANDLE

XXII

6 1/2"

2 1/4"

82

5 3/8"

13"

GENUINE
PLUMB

MARKED

SOME RUST
BUT SHOWS
LITTLE USE

FLAT ONE
SIDE.

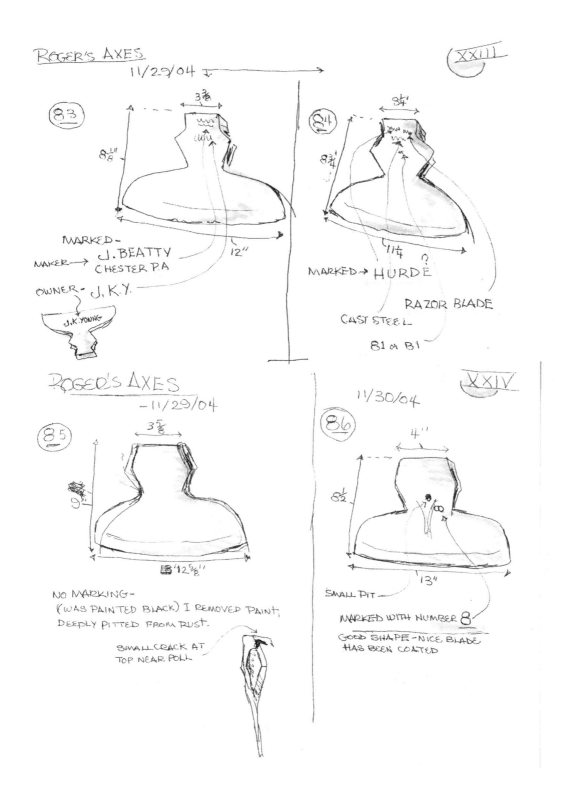

ROGER'S AXES.
11/29/04

XXIII

83

3⅞

8⅛"

MARKED-
MAKER → J. BEATTY
CHESTER P.A

OWNER - J.K.Y.

12"

J.K.YOUNG

84

3¼"

8¾"

MARKED→ HURDE

11¼

RAZOR BLADE

CAST STEEL

B1 or B1

ROGER'S AXES
-11/29/04

85

3⅝

9"

12⅝"

NO MARKING-
(WAS PAINTED BLACK) I REMOVED PAINT,
DEEPLY PITTED FROM RUST.

SMALL CRACK AT
TOP NEAR POLL

11/30/04

86

4"

8½"

SMALL PIT

13"

MARKED WITH NUMBER 8

GOOD SHAPE - NICE BLADE
HAS BEEN COATED

ROGER'S AXES
12/13/04

XXV
1/3/05

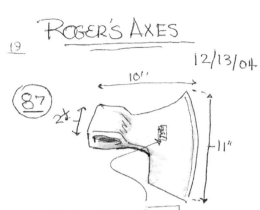

87

10"

2"

11"

MARKED - EMBOS TSG
EXCELLENT SHAPE

OLD BIG EUROPEAN BROADAXE
FROM AUSTRIA (EARLY 1900'S)

(?) IS THIS A 20TH CENTURY COPY?

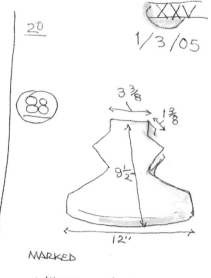

88

3 3/8

1 3/8

9 1/2"

12"

MARKED

VINTAGE AXE HEAD

VERY RUSTED & PITTED

ROGER'S AXES
1/10/05

XXVI

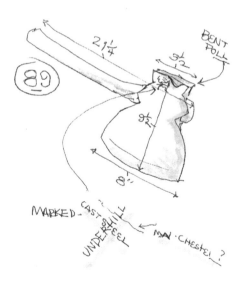

89

2 1/4

9 1/2"

BENT POLL

9 1/2"

8"

MARKED - CAST STEEL
UNDER STEEL

MAN CHESTER?

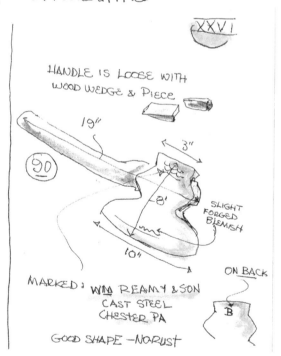

HANDLE IS LOOSE WITH
WOOD WEDGE & PIECE

90

19"

3"

8'

SLIGHT
FORGED
BLEMISH

10"

ON BACK

B

MARKED: WM REAMY & SON
CAST STEEL
CHESTER PA

GOOD SHAPE - NO RUST

1/20/05

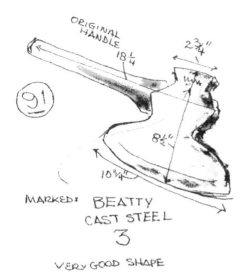

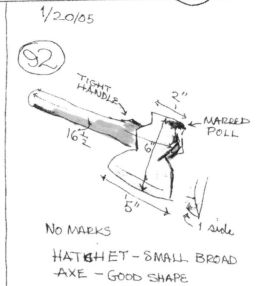

⑨1

ORIGINAL HANDLE
18¼
2¾"
8½"
10¾"

⑨2

TIGHT HANDLE
16½
2"
6"
5"
MARRED POLL
1 side

MARKED: BEATTY
CAST STEEL
3

VERY GOOD SHAPE

NO MARKS

HATCHET - SMALL BROAD
AXE - GOOD SHAPE

ROGER'S AXES

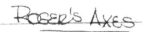

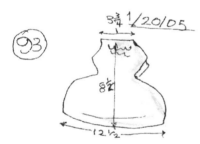

⑨3

8¾ 1/20/05

8½
12½

MARKED - 2 R. KING   {1845
U.S.A   CAST STEEL   {1865

GOOD SHAPE - VERY HEAVY

NOTE: NOT PETER KING.
THIS R. KING WORKED IN Ct.
_____ CONNE.

1/21/05

⑨4

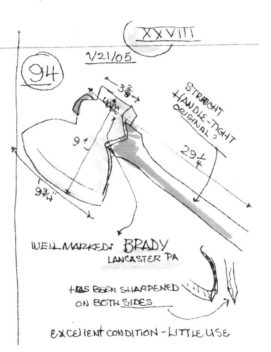

3¾
91
9¾"
STRAIGHT HANDLE - TIGHT
ORIGINAL?
29¼

WELL MARKED: BRADY
LANCASTER PA

HAS BEEN SHARPENED
ON BOTH SIDES

EXCELLENT CONDITION - LITTLE USE

1/22/05

1/22/05

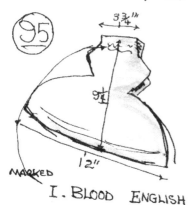

(95)

8¾"

9½"

12"

MARKED

I. BLOOD   ENGLISH
BALLSTOWN NY   CAST STEEL

GOOD SHAPE – BUT HAS BEEN SAND
BLASTED

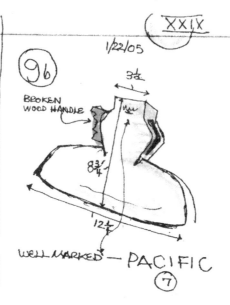

(96)

3½

BROKEN
WOOD HANDLE

8¾"

12½"

WELL MARKED – PACIFIC
(7)

VERY RUSTED & PITTED – HAS BEEN
IN THE GROUND. (BLADE – NOT POLL)

1/24/05

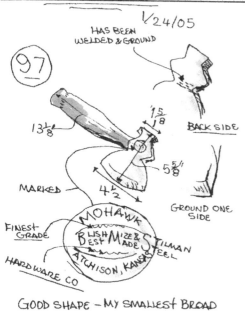

XXX

(97)

HAS BEEN
WELDED & GROUND

BACK SIDE

13⅛

1⅝
⅛

5⅝"

4½

MARKED

GROUND ONE
SIDE

FINEST
GRADE

MOHAWK
BEST MADE SIZE & STILMAN STEEL
ATCHISON, KANSAS

HARDWARE CO

GOOD SHAPE – MY SMALLEST BROAD
AXE

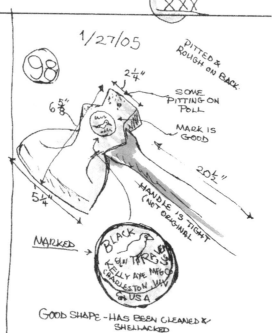

1/27/05

(98)

PITTED &
ROUGH ON BACK

2¼"

6⅝"

SOME
PITTING ON
POLL

MARK IS
GOOD

20½"

5¼"

HANDLE IS TIGHT
(NOT ORIGINAL

MARKED

BLACK AXE
KELLY AXE MFG CO
CHARLESTON W.V.
USA

GOOD SHAPE – HAS BEEN CLEANED &
SHELLACKED

2/3/05

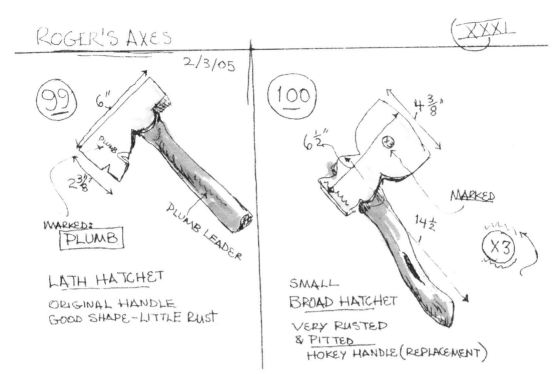

99

6"

2 3/8"

PLUMB

MARKED:
PLUMB

PLUMB LEADER

LATH HATCHET

ORIGINAL HANDLE
GOOD SHAPE - LITTLE RUST

100

4 3/8"

6 1/2"

MARKED

14 1/2

X3

SMALL
BROAD HATCHET

VERY RUSTED
& PITTED
HOKEY HANDLE (REPLACEMENT)

2/3/05

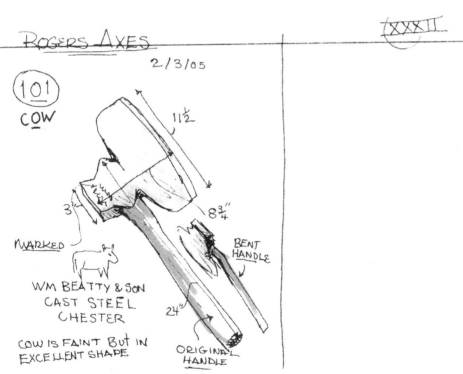

101
COW

11 1/2

3"

8 3/4"

BENT
HANDLE

MARKED

WM BEATTY & SON
CAST STEEL
CHESTER

COW IS FAINT BUT IN
EXCELLENT SHAPE

24"

ORIGINAL
HANDLE

227

2/6/05

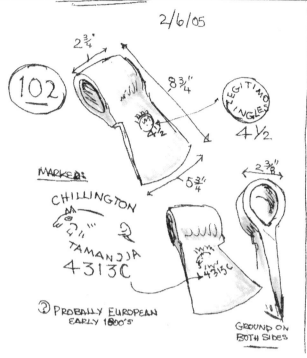

(102)

2 3/4"

8 3/4"

LEGITIMO INGLES

4 1/2

MARKED:

CHILLINGTON

TAMANJJA

4313C

② PROBALLY EUROPEAN
EARLY 1800'S

5 3/4"

2 3/8"

43135C

GROUND ON
BOTH SIDES

2/22/03

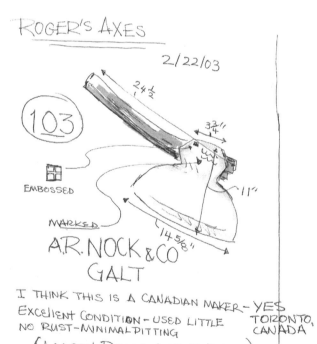

(103)

24 1/2"

3 3/4"

EMBOSSED

"11"

MARKED

14 5/8"

A.R. NOCK & CO
GALT

I THINK THIS IS A CANADIAN MAKER - YES
EXCELLENT CONDITION - USED LITTLE   TORONTO,
NO RUST - MINIMAL PITTING   CANADA

(LARGEST BROAD AXE TO DATE)

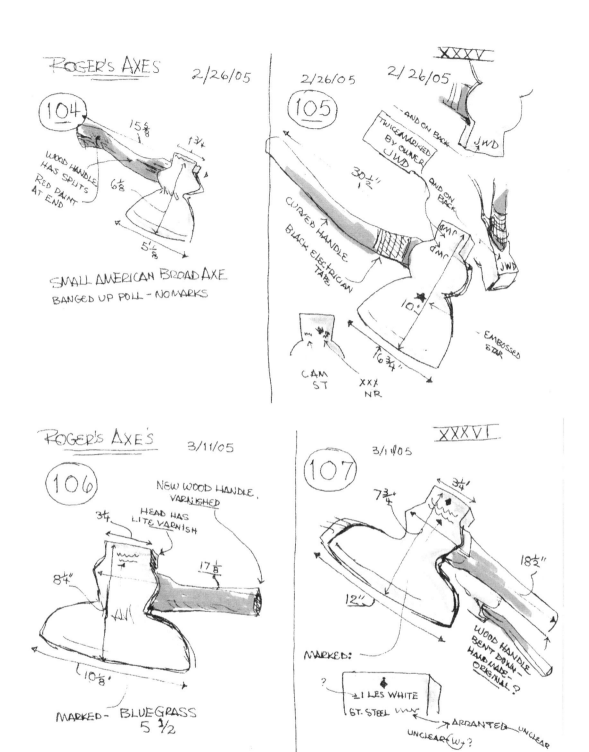

ROGER'S AXES   2/26/05   2/26/05   2/26/05

(104)   (105)
        15½"
WOOD HANDLE   TWICE MARKED
HAS SPLITS   BY OWNER
RED PAINT   JWD
AT END
        6⅛"   XXXV

                                                                    JWD
        5⅜"

SMALL AMERICAN BROAD AXE
BANGED UP POLL - NO MARKS   30½"
                                      CURVED HANDLE
                                      BLACK ELECTRICAN
                                      TAPE

                                                                    AND ON BACK

                                                            10"

                                                                    EMBOSSED
                                                                    STAR
                                      CAM
                                      ST   XXX
                                                NR
                                            16¾"

ROGER'S AXES   3/11/05   XXXVI
                                            3/11/05
(106)   (107)
        NEW WOOD HANDLE   3¼"
        VARNISHED
        HEAD HAS   7¾"
        3¼" LITE VARNISH
                    17⅛"   18½"
8¼"
                                        12"
                                                        WOOD HANDLE
        10⅛"   BENT DOWN-
                                                        HAND MADE-
MARKED - BLUEGRASS   MARKED: ORIGINAL?
        5 ½
                                        ?
                                          ±1 LES WHITE
                                          GT. STEEL   WARRANTED UNCLEAR
                                                        UNCLEAR (W?)

                                                                                    229

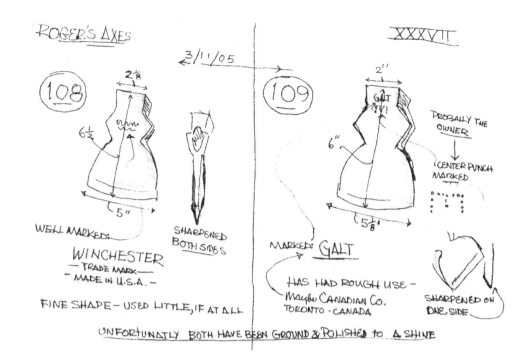

108

2⅛"

6½

WELL MARKED:

SHARPENED
BOTH SIDES

WINCHESTER
— TRADE MARK —
— MADE IN U.S.A. —

FINE SHAPE — USED LITTLE, IF AT ALL

3/11/05

109

2"

GALT

6"

5⅛"

PROBABLY THE
OWNER

CENTER PUNCH
MARKED

MARKED: GALT

HAS HAD ROUGH USE —
Maybe CANADIAN Co.
TORONTO - CANADA

SHARPENED ON
ONE SIDE

UNFORTUNATLY BOTH HAVE BEEN GROUND & POLISHED TO A SHINE

Appendix 3: Steelmaking Techniques, Terminology, and Composition

Carbon Content of Ferrous Metals

Sources vary widely in defining the *minimum* carbon content of steel, which ranges from 0.1 to 0.5% carbon. Please note the caveats that follow the definitions.*

Wrought iron: 0.01 – 0.08% carbon content (cc); soft, malleable, ductile, corrosion-resistant, and containing significant amounts of siliceous slag in bloomery produced wrought iron, with less slag in blast-furnace-derived, puddled wrought iron. Wrought iron is often noted as having ≤ 0.03% carbon content.

Malleable iron 1): 0.08 – 0.2% carbon content (cc); malleable and ductile, but harder and more durable than wrought iron; also containing significant amounts of siliceous slag in bloomery produced malleable iron, with less slag in blast furnace derived, puddled malleable iron.

Malleable iron 2): > 0.2 – 0.5% carbon content (cc). Prior to the advent of bulk-processed low carbon steel (1870), iron containing the same amount of carbon as today's "low carbon steel" (see below) was called "malleable iron." Its siliceous slag content gave it toughness and ductility, qualities not present in modern low carbon steel, hence its name. Before 1870, a wide variety of common hand and garden tools and hardware were made from malleable iron with a significantly higher carbon content than wrought iron.

Natural steel: 0.2% carbon content or greater. Natural steel containing less than 0.5% cc is synonymous with the term malleable iron. Natural steel is produced only by direct process bloomery smelting and was the only form of steel produced in Europe from the early Iron Age to the appearance of the blast furnace (1350). Small quantities of natural steel continued to be produced by bloomsmiths, especially in the bog iron furnaces of colonial New England until the late 19th century.

German steel: 0.2% carbon content or greater. Steel made from the decarburizing of cast iron in finery furnaces, as, for example, at the Saugus Ironworks after 1646. The strategy of making German steel dominated European steel production between 1400 and the advent of bulk process steel technologies, hence the term "continental method" as an alternative name for this type of steel production.

Wrought steel: 0.2 – 0.5% carbon content (cc); another name for malleable iron. Wrought steel was made from iron bar stock and was deliberately carburized during the fining process to make steel tools that are still commonplace today, such as the ubiquitous blacksmith's leg vise.

Low carbon steel: 0.2 – 0.5% carbon content (cc). Less malleable and ductile than wrought and malleable iron due to its lack of ferrosilicate, low carbon steel is harder and more durable than either and can be only slightly hardened by quenching. Some recent authors (Sherby 1995a) define low carbon steel as having 0.1% cc. Produced after 1870 as bulk process steel (e.g. by the Bessemer process), low carbon steel has all its siliceous slag content removed by oxidation. Before the advent of bulk process steel production, there was no such term as "low carbon steel." All iron that could not be hardened by quenching (< 0.5% cc) was known as "malleable" iron, more recently often referred to as "wrought" iron.

Tool steel: As with many forms of iron and steel, the term "tool steel" has multiple meanings. Tool steel has traditionally been known as steel with 0.5 – 2.0% carbon content (cc). Tool steel has the unique characteristic that it can be hardened by quenching, which then requires tempering to alleviate its brittleness. Increasing carbon content decreases the malleability of steel. If containing >1.5% carbon content, steel is not malleable, and, thus, not forgeable, at any temperature. Such steel is now called ultra high carbon steel (UHCS). Palmer, in *Tool Steel Simplified*, provides this generic description of tool steel: "Any steel that is used for the working parts of tools" (Palmer 1937, 10). The modern definition (post 1950) of tool steel is any steel containing more than 4% of one or more alloys.

Ultra high carbon steel (UHCS): 1.5 – 2.5% carbon content (cc); a modern form of hardened steel characterized by superplasticity at high temperatures and used in industrial applications, such as jet engine turbine manufacturing, where extreme strength, durability, and exact alloy content are necessary. Powdered metallurgy technology is frequently used to make UHCS.

Cast iron: 2.0 – 4.5% carbon content (cc); hard and brittle; not machinable unless annealed to produce malleable cast iron. For a more detailed description of the many varieties of cast iron see cast iron.

*Caveats to carbon content of ferrous metals

- Both modern and antiquarian sources vary widely in their definitions of wrought iron, malleable iron, and steel. Modern sources variously define steel and/or low carbon steel as iron having a carbon content greater than 0.08%, 0.1%, 0.2%, and 0.3%.

- Before the advent of bulk process steel industries (1870), which produced huge quantities of low carbon steel that could have a carbon content in the range of 0.08 – 0.5%, iron having a carbon content of < 0.5% cc was called malleable iron. Other generic terms for iron that could not be hardened by quenching (> 0.5% cc) were bar iron, wrought iron, and merchant bar.

- The 1911 edition of the *Encyclopedia Britannica* defines wrought iron as containing less than 0.3% carbon, cast iron as having 2.2% or more carbon content and steel as having an intermediate carbon content > 0.3% and < 2.2%.

- Gordon (1996) defines steel as having a carbon content > 0.2%. This cutoff point is probably the most appropriate to use in defining steel, but also poses a problem since most sources define wrought iron as having < 0.08% cc; therefore, leading to the confusion of iron with a carbon content > 0.08% but < 0.2% as being either wrought iron, low carbon steel or an orphan form of undefined iron.

- In view of the long tradition of the use of the term malleable iron, this glossary resurrects the use of that term to cover this gray area of the carbon content of ferrous metals.

A Guide to the Metallurgy of the Edge Tools at the Davistown Museum
Steelmaking Strategies 1900 BCE – 1930 CE

1. Natural Steel: 1900 BCE – 1930 CE

Natural steel was made in direct process bloomeries, either deliberately or accidentally, in the form of occasional nodules of steel (+/- 0.5% carbon content (cc)) entrained in wrought iron loups. Bloomsmiths deliberately made natural steel for sword cutlers by altering the fuel to ore ratio in the smelting process, producing heterogeneous blooms of malleable iron (0.08 to 0.2% cc) and/or natural steel (0.2 to 0.5 cc and higher) or by carburizing bar or sheet iron submerged in a charcoal fire. Manganese-laced rock ores (e.g. from Styria in Austria or from the Weald in Sussex, England) facilitated natural steel production. As a slag constituent, manganese lowered the melting temperature of slag, facilitating the more uniform uptake of carbon in the smelted iron. The Chalybeans produced the first documented natural steel at the height of the Bronze Age in 1900 BCE, using the self-fluxing iron sands from the south shores of the Black Sea. Occasional production of bloomery-derived natural steel edge tools continued in isolated rural areas of Europe and North America into the early 20th century.

2. German Steel: 1350 - 1900

German steel was produced by decarburizing blast-furnace-derived cast iron in a finery furnace, and, after 1835, in a puddling furnace. German steel tools are often molded, forged, or cast entirely of steel, as exemplified by trade and felling axes without an inserted (welded) steel bit. Such tools were a precursor of modern cast steel axes and rolled cast steel timber framing tools. German steel shared the world market for steel with English blister and crucible steel until the mid-19th century.

3. Blister Steel: 1650 - 1900

Blister steel was produced by carburizing wrought iron bar stock in a sandstone cementation furnace that protected the ore from contact with burning fuel. It was often refined by piling, hammering, and reforging it into higher quality shear or double shear steel or broken up and remelted in crucibles to make cast steel. Blister steel was often used for "steeling" (welding on a steel cutting edge or bit) on axes and other edge tools.

4. Shear Steel: 1700 - 1900

Shear steel was made from refined, reforged blister steel and used for "steeling" high quality edge tools, such as broad axes, adzes, and chisels, especially by American edge toolmakers who did not have access to, or did not want to purchase, expensive imported English cast steel. The use of shear steel was an alternative to imported English cast steel for making edge tools in America from the late 18th century to the mid-19th century.

5. Crucible Cast Steel: 1750 - 1930

Crucible cast steel was made from broken up pieces of blister steel bar stock, which was inserted into clay crucibles with small quantities of carboniferous materials (e.g. charcoal powder). After melting at high temperatures, crucible cast steel was produced in 5 to 25 kg batches and considered to be the best steel available for edge tool, knife, razor, and watch spring production. Due to lack of heat resistant clay crucibles, extensive production of high quality crucible cast steel didn't begin in the United States until after the Civil War.

6. Brescian Steel: 1350 - 1900?

Brescian Steel was a common Renaissance era strategy used in southern Europe to make, for example, steel for the condottiers of the Italian city states. Wrought or malleable iron bar stock was submerged and, thus, carburized in a bath of molten pig iron. Brescian steel cannot be visually differentiated from German steel or puddled steel, both of which were produced from decarburizing pig iron.

7. Bulk Processed Steel: 1870 f.

After the American Civil War, a number of new strategies were invented for producing large quantities of steel, especially low carbon steel, which was required by the rapid growth of the industrial age and its factory system of mass production. The first important innovation was Henry Bessemer's single step hot air blast process, followed by several variations of the Siemens-Martin open-hearth furnace and electric arc furnaces. For edge tool production, the electric arc furnace supplanted, and then replaced, crucible cast steel in the early decades of the 20th century. A few modern drop-forged edge tools are included in this exhibition as examples of modern bulk process steel producing strategies.

Edge Toolmaking Techniques 1900 BCE – 1930 CE

Shaping and Forging by Hand

A. Forge welding: Edge carburizing by heating, followed by hammering and additional heat treatment

B. Steeling: The welding of a steel bit onto an iron shaft or body

C. Pattern-welding: The welding together of alternating layers of sheet iron and steel, used by knife and swordmakers; seldom used by edge toolmakers

D. Molding: The shaping of short lengths of hot malleable iron or German steel bar stock in an iron pattern; sometimes the iron pattern was water-cooled. This method was not used after blister steel became widely available around 1700.

Shaping and Forging by Machine

E. Rolling: The hot rolling of cast steel into bar stock, and its further shaping by the formation of sockets, grinding, and further forging, both before and after additional thermal treatment.

F. Casting: The hot rolling of cast steel into steel bar stock compatible with its further shaping in molds or patterns by drop-forging, as in the drop-forging of cast steel axes.

G. Drop-forging: The hydraulic pressing of low carbon steel and malleable iron into tool forms by using dies as patterns as in the mass production of factory-made hand tools. Also, its casting in special purpose molds for the production of machinery and equipment of every conceivable use.

Most hand tools made in the 20th century show no evidence of hand work, but, in a minority of cases, (e.g. the ax) there is no clear distinction between the hand-forged and the machine-made tool until the late 20th century. Most edge tools made before 1930 are "hand-forged" or "forge welded" to some extent, no matter the technique used to "steel" their edges. The trip hammer and the water wheel are examples of machines that assisted edge toolmakers in the forging of their tools. The advent of the modern rolling mill (Henry Cort, 1784) for hot rolling cast steel bar stock did not end the long tradition of hand-forging an edge tool. When the Collins Ax Factory began drop-forging all steel axes sometime after 1837, many smaller ax companies continued hand-forging and hand hammering axes they produced, often with the aid of other machinery, well into the 20th century. The evolution from hand-forging to machine forging (drop-forging) hand tools was thus a gradual process. One goal of the creative economy of the post-industrial era is the revival of handmade hand toolmaking strategies and techniques.

Classification for Metals and Alloys

Unified Numbering System (UNS) for Metals and Alloys. From Oberg et al. ([1914] 1996, 402) *Machinery's Handbook: 25th edition: A reference book for the mechanical engineer, designer, manufacturing engineer, draftsman, toolmaker, and machinist.*

UNS Series	Metal
A00001 to A99999	Aluminum and aluminum alloys
C00001 to C99999	Copper and copper alloys
D00001 to D99999	Specified mechanical property steels
E00001 to E99999	Rare earth and rare earthlike metals and alloys
F00001 to F99999	Cast irons
G00001 to G99999	AISI and SAE carbon and alloy steels (except tool steels)
H00001 to H99999	AISI and SAE H-steels
J00001 to J99999	Cast steels (except tool steels)
K00001 to K99999	Miscellaneous steels and ferrous alloys
L00001 to L99999	Low-melting metals and alloys
M00001 to M99999	Miscellaneous nonferrous metals and alloys
N00001 to N99999	Nickel and nickel alloys
P00001 to P99999	Precious metals and alloys
R00001 to R99999	Reactive and refractory metals and alloys
S00001 to S99999	Heat and corrosion resistant (stainless) steels
T00001 to T99999	Tool steels, wrought and cast
W00001 to W99999	Welding filler metals
Z00001 to Z99999	Zinc and zinc alloys

Compositions of AISI-SAE Standard Carbon Steels

From Oberg et al. ([1914] 1996, 407-8) *Machinery's Handbook: 25th edition: A reference book for the mechanical engineer, designer, manufacturing engineer, draftsman, toolmaker, and machinist.*

| AISI NO. | Composition (%) | | | | SAE No. |
	C	Mn	P Max.	S Max.	
Nonresulfurized Grades – 1 per cent Mn (max)					
1005	0.06 max	.035 max	.040	.050	1005
1006	0.08 max	0.25-0.40	.040	.050	1006
1008	0.10 max	1.30-0-50	.040	.050	1008
1010	0.08-0.13	0.30-0.60	.040	.050	1010
1011	0.08-0.13	0.60-0.90	.040	.050	—
1012	0.10-0.15	0.30-0.60	.040	.050	1012
1013	0.11-0.16	0.50-0.80	.040	.050	1013
1015	0.13-0.18	0.30-0.60	.040	.050	1015
1016	0.13-0.18	0.60-0.90	.040	.050	1016
1017	0.15-0.20	0.30-0.60	.040	.050	1017
1018	0.15-0.20	0.60-0.90	.040	.050	1018
1019	0.15-0.20	0.70-1.00	.040	.050	1019
1020	0.10-0.23	0.30-0.60	.040	.050	1020
M1020	0.17-0.24	0.25-0.60	.040	.050	—
1021	0.18-0.25	0.60-0.90	.040	.050	1021
1022	0.18-0.23	0.70-1.00	.040	.050	1022
1023	0.20-0.25	0.30-0.60	.040	.050	1023
1025	0.22-0.28	0.30-0.60	.040	.050	1025
1026	0.22-0.28	0.60-0.90	.040	.050	1026
1029	0.25-0.31	0.60-0.90	.040	.050	—
1030	0.28-0.34	0.60-0.90	.040	.050	1030
1034	0.32-0.28	0.50-0.80	.040	.050	—
1035	0.32-0.38	0.60-0.90	.040	.050	1035
1037	0.32-0.38	0.70-1.00	.040	.050	1037
1038	0.35-0.42	0.60-0.90	.040	.050	1038

AISI NO.	Composition (%)				SAE No.
	C	Mn	P Max.	S Max.	
1039	0.37-0.44	0.70-1.00	.040	.050	1039
1040	0.37-0.44	0.60-0.90	.040	.050	1040
1042	0.40-0.47	0.60-0.90	.040	.050	1042
1043	0.40-0.47	0.70-1.00	.040	.050	1043
1044	0.43-0.50	0.30-0.60	.040	.050	1044
M1044	0.40-0.50	0.25-0.60	.040	.050	—
1045	0.43-0.50	0.60-0.90	.040	.050	1045
1046	0.43-0.50	0.70-1.00	.040	.050	1046
1049	0.46-0.53	0.60-0.90	.040	.050	1049
1050	0.48-0.55	0.60-0.90	.040	.050	1050
M1053	0.48-0.55	0.70-1.00	.040	.050	—
1055	0.50-0.60	0.60-0.90	.040	.050	1055
1059	0.55-0.65	0.50-0.80	.040	.050	1059
1060	0.55-0.65	0.60-0.90	.040	.050	1060
1064	0.60-0.70	0.50-0.80	.040	.050	1064
1065	0.60-0.70	0.60-0.90	.040	.050	1065
1069	0.65-0.75	0.40-0.70	.040	.050	1069
1070	0.65-0.75	0.60-0.90	.040	.050	1070
1071	0.65-0.70	0.75-1.05	.040	.050	—
1074	0.70-0.80	0.50-0.80	.040	.050	1074
1075	0.70-0.80	0.40-0.70	.040	.050	1075
1078	0.72-0.85	0.30-0.60	.040	.050	1078
1080	0.75-0.88	0.60-0.90	.040	.050	1080
1084	0.80-0.93	0.60-0.90	.040	.050	1084
1086	0.80-0.93	0.30-0.50	.040	.050	1086
1090	0.85-0.98	0.60-0.90	.040	.050	1090
1095	0.90-1.03	0.30-0.50	.040	.050	1095
Nonresulferized Grades – Over 1 per cent Mn					
1513	0.10-0.16	1.10-1.40	.040	.050	1513
1518	0.15-0.21	1.10-1.40	.040	.050	—

AISI NO.	Composition (%)				SAE No.
	C	Mn	P Max.	S Max.	
1522	0.18-0.24	1.10-1.40	.040	.050	1522
1524	0.19-0.25	1.35-1.65	.040	.050	1524
1525	0.23-0.29	0.80-1.10	.040	.050	------
1526	0.22-0.29	1.10-1.40	.040	.050	1526
1527	0.22-0.29	1.20-1.50	.040	.050	1527
1536	0.30-0.37	1.20-1.50	.040	.050	1536
1541	0.36-0.44	1.35-1.65	.040	.050	1541
1547	0.43-0.51	1.35-1.65	.040	.050	-----
1548	0.44-0.52	1.10-1.40	.040	.050	1548
1551	0.45-0.56	0.85-1.15	.040	.050	1551
1552	0.47-0.55	1.20-1.50	.040	.050	1552
1561	0.55-0.65	0.75-1.05	.040	.050	1561
1566	0.60-0.71	0.85-1.15	.040	.050	1566
1572	0.65-0.76	1.00-1.30	.040	.050	-----
Free-Machining Grades - Resulfurized					
1108	0.08-0.13	0.50-0.80	.040	0.08-0.13	1108
1109	0.08-0.13	0.60-0.90	.040	0.08-0.13	1109
1110	0.08-0.13	0.30-0.60	.040	0.08-0.13	1110
1116	0.14-0.20	1.10-1.40	.040	0.16-0.23	1116
1117	0.14-0.20	1.00-1.30	.040	0.08-0.13	1117
1118	0.14-0.20	1.30-1.60	.040	0.08-0.13	1118
1119	0.14-0.20	1.0-1.30	.040	0.24-0.33	1119
1132	0.27-0.34	1.35-1.65	.040	0.08-0.13	1132
1137	0.32-0.39	1.35-1.65	.040	0.08-0.13	1137
1139	0.35-0.43	1.35-1.65	.040	0.13-0.20	1139
1140	0.37-0.44	0.70-1.00	.040	0.08-0.13	1140
1141	0.37-0.45	1.35-1.65	.040	0.08-0.13	1141
1144	0.40-0.48	1.35-1.65	.040	0.24-0.33	1144
1145	0.42-0.49	0.70-1.00	.040	0.04-0.07	1145
1146	042-0.49	0.70-1.00	.040	0.08-0.13	1146

	Composition (%)				
AISI NO.	C	Mn	P Max.	S Max.	SAE No.
1151	0.48-0.55	0.70-1.00	.040	0.08-0.13	1151
Free-Machining Grades – Resulfurized and Rephosphorized					
1211	0.13 max	0.60-0.90	0.07-0.12	0.10-0.15	1211
1212	0.13 max	0.70-1.00	0.07-0.12	0.16-0.23	1212
1213	0.13 max	0.70-1.00	0.07-0.12	0.24-0.33	1213
1215	0.09 max	0.75-1.05	0.04-0.09	0.26-0.35	1215
12L 14	0.15 max	0.85-1.15	0.04-0.09	0.26-0.35	12L14

Classification for Carbon and Alloy Steels

AISI-SAE System of Designating Carbon and Alloy Steels. From Oberg et al. ([1914] 1996, 406) *Machinery's Handbook: 25th edition: A reference book for the mechanical engineer, designer, manufacturing engineer, draftsman, toolmaker, and machinist.*

SAE designation	Type
Carbon and alloy steel grades	
Carbon steels	
10xx	Plain carbon (Mn 1.00% max)
11xx	Resulfurized
12xx	Resulfurized and rephosphorized
15xx	Plain carbon (Mn 1.00% to 1.65%)
Manganese steels	
13xx	Mn 1.75%
Nickel steels	
23xx	Ni 3.50%
25xx	Ni 5.00%
Nickel-chromium steels	
31xx	Ni 1.25%, Cr 0.65% or 0.80%
32xx	Ni 1.25%, Cr 1.07%
33xx	Ni 3.50%, Cr 1.50% or 1.57%
34xx	Ni 3.00%, Cr 0.77%
Molybdenum steels	
40xx	Mo 0.20% or 0.25% or 0.25% Mo & 0.042 S
44xx	Mo 0.40% or 0.52%
Chromium-molybdenum (Chromoly) steels	
41xx	Cr 0.50% or 0.80% or 0.95%, Mo 0.12% or 0.20% or 0.25% or 0.30%
Nickel-chromium-molybdenum steels	
43xx	Ni 1.82%, Cr 0.50% to 0.80%, Mo 0.25%
43BVxx	Ni 1.82%, Cr 0.50%, Mo 0.12% or 0.35%, V 0.03% min
47xx	Ni 1.05%, Cr 0.45%, Mo 0.20% or 0.35%
81xx	Ni 0.30%, Cr 0.40%, Mo 0.12%
81Bxx	Ni 0.30%, Cr 0.45%, Mo 0.12%
86xx	Ni 0.55%, Cr 0.50%, Mo 0.20%
87xx	Ni 0.55%, Cr 0.50%, Mo 0.25%
88xx	Ni 0.55%, Cr 0.50%, Mo 0.35%
93xx	Ni 3.25%, Cr 1.20%, Mo 0.12%
94xx	Ni 0.45%, Cr 0.40%, Mo 0.12%
97xx	Ni 0.55%, Cr 0.20%, Mo 0.20%
98xx	Ni 1.00%, Cr 0.80%, Mo 0.25%
Nickel-molybdenum steels	
46xx	Ni 0.85% or 1.82%, Mo 0.20% or 0.25%

Carbon and alloy steel grades	
SAE designation	**Type**
48xx	Ni 3.50%, Mo 0.25%
Chromium steels	
50xx	Cr 0.27% or 0.40% or 0.50% or 0.65%
50xxx	Cr 0.50%, C 1.00% min
50Bxx	Cr 0.28% or 0.50%
51xx	Cr 0.80% or 0.87% or 0.92% or 1.00% or 1.05%
51xxx	Cr 1.02%, C 1.00% min
51Bxx	Cr 0.80%
52xxx	Cr 1.45%, C 1.00% min
Chromium-vanadium steels	
61xx	Cr 0.60% or 0.80% or 0.95%, V 0.10% or 0.15% min
Tungsten-chromium steels	
72xx	W 1.75%, Cr 0.75%
Silicon-manganese steels	
92xx	Si 1.40% or 2.00%, Mn 0.65% or 0.82% or 0.85%, Cr 0.00% or 0.65%
High-strength low-alloy steels	
9xx	Various SAE grades
xxBxx	Boron steels
xxLxx	Leaded steels

Classification of Tool Steels

From Oberg et al. ([1914] 1996, 454) *Machinery's Handbook: 25th edition: A reference book for the mechanical engineer, designer, manufacturing engineer, draftsman, toolmaker, and machinist.*

Category Designation	Letter Symbol	Group Designation
High-Speed Tool Steels	M	Molybdenum types
	T	Tungsten types
Hot-Work Tool Steels	H_1-H_{19}	Chromium types
	H_{20}-H_{39}	Tungsten types
	H_{40}-H_{59}	Molybdenum types
Cold-Work Tool Steels	D	High-carbon, high-chromium types
	A	Medium-alloy, air-hardening types
	O	Oil-hardening types
Shock-Resisting Tool Steels	S	…
Mold Steels	P	…
Special-Purpose Tool Steels	L	Low-alloy types
	F	Carbon-tungsten types
Water-Hardening Tool Steels	W	…

Bibliography

Asimov, Isaac. 1988. *Understanding physics: 3 volumes in one*. NY: Dorset Press.

Bain, Edgar C. 1939. *Functions of the alloying elements in steel*. Cleveland, OH: American Society for Metals.

Baird, Ron and Comerford, Dan. 1989. *The hammer the king of tools: A collectors handbook*. Fair Grove, MO: Self-published.

Bealer, Alex W. 1976. *The art of blacksmithing*. Edison, NJ: Castle Books.

Rice, Franklin Pierce. (1899). *Worcester of eighteen hundred and ninety-eight: Fifty years a city: A graphic representation of its institutions, industries, and leaders*. Worcester, MA: F.S. Blanchard.

Brack, H.G. 2008a. *Art of the Edge Tool: The ferrous metallurgy of New England shipsmiths and toolmakers*. Hulls Cove, ME: Pennywheel Press.

Brack, H.G. 2008b. *Steel- and Toolmaking Strategies and Techniques before 1870*. Hulls Cove, ME: Pennywheel Press.

Brack, H.G. 2013. *Handbook for Ironmongers: A glossary of ferrous metallurgy terms*. Hulls Cove, ME: Pennywheel Press.

Bradford, John W. 2011. *The 1607 Popham Colony's pinnace Virginia: An in-context design of Maine's first ship*. Rockland, ME: Maine Authors Publishing.

Brain, Jeffrey Phipps. 2011. *Fort St. George XIV: 2011 excavations at the site of the 1607-1608 Popham colony on the Kennebec River in Maine*. Salem, MA: Peabody Essex Museum.

Chard, Jack. 1995. *Making iron & steel: The historic processes 1700 – 1900*. Ringwood, NJ: North Jersey Highlands Historical Society.

Day, Joan and Tylecote, R. F. Eds. 1991. *The Industrial Revolution in metals*. Brookfield, VT: The Institute of Metals.

Diderot, Denis and d'Alembert, Jean Baptiste le Rond. [1751-65] 1964. *Recueil de planches sur les sciences, les arts libéraux, et les arts mechaniques avec leur explication.* Volume 1. Paris: Au Cercle du Livre Precieux.

Diderot, Denis and d'Alembert, Jean Baptiste le Rond. [1751-65] 1965a. *Recueil de planches sur les sciences, les arts libéraux, et les arts mechaniques avec leur explication.* Volume 2. Paris: Au Cercle du Livre Precieux.

Diderot, Denis and d'Alembert, Jean Baptiste le Rond. [1751-65] 1965b. *Recueil de planches sur les sciences, les arts libéraux, et les arts mechaniques avec leur explication.* Volume 3. Paris: Au Cercle du Livre Precieux.

Diderot, Denis and d'Alembert, Jean Baptiste le Rond. [1751-65] 1965c. *Recueil de planches sur les sciences, les arts libéraux, et les arts mechaniques avec leur explication.* Volume 4. Paris: Au Cercle du Livre Precieux.

Diderot, Denis and d'Alembert, Jean Baptiste le Rond. [1751-65] 1965d. *Recueil de planches sur les sciences, les arts libéraux, et les arts mechaniques avec leur explication.* Volume 5. Paris: Au Cercle du Livre Precieux.

Diderot, Denis and d'Alembert, Jean Baptiste le Rond. [1751-65] 1966. *Recueil de planches sur les sciences, les arts libéraux, et les arts mechaniques avec leur explication.* Volume 6. Paris: Au Cercle du Livre Precieux.

Fitzhugh, William W., and Jacqueline S. Olin, eds. 1993. *Archeology of the Frobisher voyages.* Washington, DC: Smithsonian Institution Press.

Galileo, Galilei. 1960. *On Motion and On Mechanics, Comprising De Motu (ca. 1590) and Le Meccaniche (ca. 1600).* Madison, WI: University of Wisconsin Press.

Gordon, Robert. B. 1996. *American iron 1607 – 1900.* Baltimore, MD: John Hopkins University Press.

Heavrin, Charles A. 1998. *The axe and man.* Mendham, NJ: Astragal Press.

Hock, Ron. 2009. *The perfect edge: The ultimate guide to sharpening for woodworkers.* Cincinnati, OH: Popular Woodworking Books.

Jacob, Walter W. 2011. *Stanley woodworking tools: The finest years: Research and type studies adapted from The Chronicle of the Early American Industries Association.* Hebron, MD: Early American Industries Association.

Kauffman, Henry J. [1966] 1995. *Metalworking trades in early America.* Mendham, NJ: Astragal Press.

Kauffman, Henry J. 1972. *American Axes: A survey of their development and their makers.* Brattleboro, VT: The Stephen Greene Press.

Klenman, Allen. 1990. *Axe makers of North America.* Victoria, BC: Whistle Punk Books.

Martin J. Donnelly Antique Tools. (1993). *An introduction to classic American machinist tools.* Bath, NY: Martin J. Donnelly Antique Tools.

Moxon, Joseph. [1703] 1989. *Mechanick exercises or the doctrine of handiworks.* Morristown, NJ: The Astragal Press.

Murdock, Bartlett. (1937). *Blast furnaces of Carver, Plymouth County.* Poughkeepsie, NY: Self-published.

Oberg, Erik, Jones, Franklin D., Horton, Holbrook L. and Ryffel, Henry H. [1914] 1996. *Machinery's Handbook: 25th edition: A reference book for the mechanical engineer, designer, manufacturing engineer, draftsman, toolmaker, and machinist.* NY: Industrial Press Inc.

Palmer, Frank R., Luerssen, George V. and Pendleton, Joseph S., Jr. 1978. *Tool steel simplified.* Radnor, PA: Chilton Company.

Pleiner, Radomir. 1962. *Staré evropské kovářství.* Prague: Alteuropäisches Schmiedehandwerk.

Postman, Richard. 1998. *Anvils in America.* Berrian Springs, MI: Self-published.

Rees, Jane. [1787] 2004. *A directory of Sheffield: Including the manufacturers of the adjacent villages.* Wiltshire, UK: Tools & Trades History Society.

de Réaumur, René Antoine Ferchault. 1722. *L'art de convertir le fer forgé en acier.* Paris, France.

Rosenberg, Nathan. 1975. America's rise to woodworking leadership. In: Hindle, Brooke, Ed. *America's wooden age: Aspects of its early technology.* Tarrytown, NY: Sleepy Hollow Restorations.

Salaman, R.A. 1975. *Dictionary of tools used in the woodworking and allied trades, c. 1700-1975.* NY: Charles Scribner's Sons.

Sellens, Alvin. 1990. *Dictionary of American hand tools: A pictorial synopsis.* Augusta, KS: Self-Published.

Sherby, Oleg D. 1995. Damascus steel and superplasticity – Part I: Background, superplasticity, and genuine Damascus steels. *SAMPE Journal.* 31:10-7.

Sloane, Eric. 1964. *A museum of early American tools.* NY: Funk & Wagnalls.

Sloane, Eric. 2004. *Diary of an early American boy Noah Blake 1805.* NY: Dover Publications.

Smith, Joseph. [1816] 1975. *Explanation or key to the various manufactories of Sheffield, with engravings of each article.* South Burlington, VT: Early American Industries Association.

Smith, Roger K. 1981-1992. *Patented transitional & metallic planes in America 1827 - 1927.* 2 vols. North Village Publishing Co., Lancaster, MA.

Story, Dana. 1995. *The shipbuilders of Essex: A chronicle of Yankee endeavor.* Gloucester, MA: Ten Pound Island Book Company.

Timmins, R. & Sons. [no date] 1976. *Tools for the trades and crafts: An eighteenth century pattern book.* Fitzwilliam, NH: K. Roberts Pub. Co.

Tylecote, Ronald F. 1987. *The early history of metallurgy in Europe.* London: Longmans Green.

Warren, William L. July 1992. The Litchfield Iron Works in Bantam, Connecticut 1729 - 1825. *Tools & Technology.* Special Edition.

Weston, Thomas. 1906. *History of the town of Middleboro Massachusetts.* New York: Houghton, Mifflin and Company.

Winsor, Justin, Ed. 1881. *The memorial history of Boston including Suffolk County, Massachusetts. 1630-1880*. 4 vols. Boston, MA: James R. Osgood and Company.

Made in the USA
Lexington, KY
14 July 2014